ELECTRON PARAMAGNETISM

JUAN A. McMILLAN
Argonne National Laboratory

REINHOLD BOOK CORPORATION
A subsidiary of Chapman-Reinhold, Inc.

NEW YORK   AMSTERDAM   LONDON

PHYSICS

Illustrations by the author

To my wife

Preface

This book is not intended for specialists; it has been written for chemists doing research in other fields and for chemistry students. It contains material that could be offered as an advanced undergraduate course for chemists who wish to specialize in the field of solid state. It is presumed that the reader has been exposed to basic electromagnetism and thermodynamics as well as to atomic theory. Only a minimum background in elementary calculus, vector algebra and trigonometry has been assumed. For this reason it has been necessary to develop the mathematical treatment behind the method whenever needed. Such an approach has the advantage that the mathematical treatment and the physical phenomenon illuminate each other. Unfortunately, such treatments are necessarily lengthy, and some demanding readers may not find enough advanced information. For this purpose, general references are provided.

If a book of this type is to succeed, it must not contain loose statements or too sophisticated arguments. It must appeal to advanced undergraduate and graduate students, as well as to teachers and chemists already active in research. Whether or not I have succeeded in this task is a decision that must be left to the reader.

It is my pleasure to acknowledge the interest of Dr. Bernard Smaller of Argonne National Laboratory, to whom I owe many enlightening discussions, and to Dr. Teodoro Halpern of the University of Chicago, who was patient in reading and kind in criticizing the manuscript.

J. A. M.

Argonne, 1967

Contents

Contents x

PART I

Theory

1 *Magnetic polarization*

1.1 Electromagnetic field

There are four basic vectors used for describing the fundamental properties of the electromagnetic field: two are electric (**E** and **D**) and two are
magnetic (**H** and **B**). The first one of each set is called *intensity*, the
second, *induction*. The electric vector **D** is also called the *electric displacement vector*, for it accounts for the Maxwell displacement current inside
a dielectric. Electric vectors are defined by using properties that do not
depend on the medium, as follows:

The electric intensity **E** is defined by

$$dF = E dq \qquad (1-1)$$

where dF is the force that acts on a positive point charge, dq.

The electric induction **D** is defined by

$$dq = (1/4\pi) D_n d\sigma \qquad (1-2)$$

where dq is the charge induced in the element of area, $d\sigma$, and D_n is the component of **D** in the direction normal to $d\sigma$. Eq. (1-2) is valid in sign if dq is the charge induced in the outside of $d\sigma$, and the normal to $d\sigma$ is positive outward. Vectors **E** and **D** are related by

$$\mathbf{D} = \boldsymbol{\epsilon} \cdot \mathbf{E} \qquad (1\text{-}3)$$

where ϵ is a second-rank tensor characterizing the medium, known as *dielectric polarizability tensor*. In isotropic media, ϵ reduces to the scalar dielectric constant ϵ and Eq. (1-3) becomes

$$\mathbf{D} = \epsilon\mathbf{E} \qquad (1\text{-}4)$$

The magnetic vectors are defined as follows:

The torque **T** that acts upon a linear dipole of magnetic moment **m** in a magnetic field **B** is

$$\mathbf{T} = \mathbf{m} \times \mathbf{B} \qquad (1\text{-}5)$$

where the sign $\times$ indicates the vector product in a right-handed system, as shown in Fig. 1-1. The right-handed system is used throughout the book.

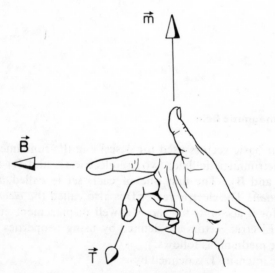

FIG. 1.1 Right-handed vector product.

The vector **B** also accounts for the electromotive force (emf) induced by its time variation:

$$\oint \mathbf{E} \cdot d\boldsymbol{\ell} = -(1/c)(d/dt) \int_\Sigma B_n d\sigma \qquad (1\text{-}6)$$

where the left-hand term is the emf induced in a loop and the right-hand term is integrated over any area limited by the loop. B_n is the component of **B** along the direction normal to $d\sigma$. Eq. (1-6) may also be written

$$\text{emf} = -(1/c)(d/dt)\Phi \tag{1-7}$$

where

$$\Phi = \int_{\Sigma} B_n d\sigma \tag{1-8}$$

is the total magnetic flux. The meaning of Eqs. (1-6) and (1-7) is that the emf is positive when the magnetic flux decreases, and conversely, which is true if the sense of integration and the sense of the normal in Eq. (1-6) are related by the right-handed screw rule.

The vector **H** is related to **B** by

$$\mathbf{B} = \boldsymbol{\mu} \cdot \mathbf{H} \tag{1-9}$$

where $\boldsymbol{\mu}$ is a second-rank tensor characterizing the medium, known as the *magnetic permeability tensor*. In isotropic media, $\boldsymbol{\mu}$ reduces to the scalar permeability μ and equation (1-9) becomes

$$\mathbf{B} = \mu\mathbf{H} \tag{1-10}$$

Eqs. (1-3) and (1-9), as well as their isotropic versions (1-4) and (1-10), are similar in many respects and treated in a similar way. The isotropic electric expression may be put in the form

$$\mathbf{D} = \epsilon\mathbf{E} = \mathbf{E} + 4\pi\mathbf{P} = \mathbf{E}(1 + 4\pi\kappa_E) \tag{1-11}$$

where **P** is the electric polarization and

$$\kappa_E = P/E \tag{1-12}$$

is the electric susceptibility per unit volume. Similarly, the isotropic magnetic expression becomes

$$\mathbf{B} = \mu\mathbf{H} = \mathbf{H} + 4\pi\mathbf{M} = \mathbf{H}(1 + 4\pi\kappa) \tag{1-13}$$

where **M** is the magnetic polarization and

$$\kappa = M/H \tag{1-14}$$

is the magnetic susceptibility per unit volume.

1.2 Systems of units

A nonrationalized system of units is used in which 4π appears in Eqs. (1-11) and (1-13). The reason is that most of the results in the literature today are expressed in units of κ, as defined by Eq. (1-13). In choosing a rationalized system by defining

$$\mathbf{B} = \mathbf{H} + \mathbf{M} \tag{1-15}$$

the experimental results would have to be multiplied by 4π. Another problem appears with the choice of the system of units (i.e., electromagnetic CGS or international MKSA). In the field of paramagnetism, the practice is still to use the CGS system. It seems that—in this particular field—the MKSA system does not have any obvious advantage. The reader will find that cm^{-1} is still used as the unit of wave number (with the connotation of energy), gauss as the unit of magnetic field (**B**), and 10^{-6} electromagnetic units CGS (emu) as the unit of magnetic susceptibility. This is an epoch of transition in regards to systems of units, and there is still some controversy about the advantages of a rationalized electromagnetic system. In view of the overwhelming amount of data already published using the CGS system, it would seem unwise, in an introductory book, to increase the complexity of the subject by incorporating the difficulties arising from the use of the rationalized MKSA system. The reader should, however, be aware of the fact that he will find, among the general references, works in which both systems, MKSA or CGS, are used. In the latter case, following the system developed by Gauss, all electric quantities are measured in electrostatic units (esu) and all magnetic quantities in emu units, even when both kinds appear in the same equation. The bridge between both systems is c, the velocity of light in a vacuum, which appears on one side of the equation. (See, for example, Eqs. (1-6) and (1-7).) Clearly, the velocity of light is absent in purely magnetic and purely electric equations. An additional problem arises from the indiscriminate use of **H** and **B**, which is discussed in Section 1.7.

1.3   Energy of the magnetic field

If a linear relation exists between **H** and **B** (diamagnetic and paramagnetic substances), the energy per unit volume of the magnetic field is

$$W = \frac{\mathbf{H} \cdot \mathbf{B}}{8\pi} \tag{1-16}$$

Since, in the general anisotropic case, **B** is given by

$$\mathbf{B} = (1 + 4\pi\kappa \cdot)\mathbf{H} \tag{1-17}$$

Eq. (1-16) may then be written

$$W = \frac{\mathbf{H} \cdot (1 + 4\pi\kappa \cdot)\mathbf{H}}{8\pi} \tag{1-18}$$

In a vacuum ($\kappa = 0$) the magnetic energy is given by

$$W_{\text{vac}} = H^2/8\pi \tag{1-19}$$

The energy of magnetized matter, after subtracting Eq. (1-19), results in

$$W_m = \frac{1}{2}(\mathbf{H} \cdot \kappa \cdot \mathbf{H}) \tag{1-20}$$

and, for isotropic media

$$W_m = \frac{1}{2}\kappa H^2 \tag{1-21}$$

The volume susceptibility κ is often replaced by the mass susceptibility

$$\chi = \kappa/\rho \tag{1-22}$$

where ρ is the density of the substance. The susceptibility of interest in magnetochemistry is the susceptibility referred to one gram molecule, one gram atom, or one gram ion. It is obtained by multiplying the susceptibility χ of Eq. (1-22) by the molecular weight, the atomic weight, or the weight of one gram ion, respectively. Eqs. (1-20) and (1-21) are important in magnetochemistry, for they provide the theoretical basis of the magnetogravimetric methods studied in Chapters 6 and 7.

1.4 Magnetic moment and angular momentum of an electric charge

When a body is moving in a central field of forces, its angular momentum due to the motion

$$\mathbf{G} = m_o\mathbf{v} \times \mathbf{r} \tag{1-23}$$

is constant in the absence of external torque. The quantities m_o, $\mathbf{v}$ and $\mathbf{r}$ are the mass, the velocity and the position radius vector. If the body is charged, the angular motion will create a magnetic moment

$$\mathbf{m} = \int i\,d\sigma \tag{1-24}$$

where $d\sigma$ is an element of the area enclosed by the orbit and i is the equivalent current. The integral of Eq. (1-24) clearly includes orbits of any shape. One can substitute $d\sigma$ by

$$d\sigma = \frac{1}{2}d\ell \times \mathbf{r} \tag{1-25}$$

where $d\ell$ is an element of the orbit and $\mathbf{r}$ the radius vector referred to above. Since

$$i\,d\ell = (dq/dt)\mathbf{v}\,dt = \mathbf{v}\,dq \tag{1-26}$$

Eq. (1-24) becomes, eliminating i and $d\sigma$,

$$\mathbf{m} = \frac{1}{2} \int (\mathbf{v} \times \mathbf{r}) \, dq = \frac{1}{2} q (\mathbf{v} \times \mathbf{r}) \tag{1-27}$$

Now, since the angular momentum

$$\mathbf{G} = m_o (\mathbf{v} \times \mathbf{r}) \tag{1-23}$$

is constant, Eq. (1-27) results in

$$\mathbf{m} = \gamma \, \mathbf{G} \tag{1-28}$$

where

$$\gamma = (q/2m_o) \tag{1-29}$$

is known as the magnetogyric ratio.

1.5 Precession theorem

When an atom with a permanent magnetic moment $\mathbf{m}$ is placed in a steady magnetic field $\mathbf{B}$, a torque

$$\mathbf{T} = \mathbf{m} \times \mathbf{B} \tag{1-5}$$

acts upon it. The angular momentum $\mathbf{G}$ then changes at a rate equal to this torque, i.e.,

$$(d/dt)\mathbf{G} = \mathbf{m} \times \mathbf{B} \tag{1-30}$$

Since $\mathbf{G}$ is parallel to $\mathbf{m}$, the vector product $(\mathbf{m} \times \mathbf{B})$ is perpendicular to $\mathbf{G}$. The meaning of $(d/dt)\mathbf{G}$ is then clear: the vector $\mathbf{G}$ rotates but remains unchanged in absolute value.

Introduction of Eq. (1-28) in (1-30) leads to

$$(d/dt)\mathbf{G} = \gamma \mathbf{G} \times \mathbf{B} \tag{1-31}$$

In order to solve this vector equation, one writes the components in an orthogonal set of coordinates—chosen in such a way that $B_x = B_y = 0$, (i.e., $B_z = B$). Then

$$(d/dt)G_x = \gamma B G_y \tag{1-32}$$

$$(d/dt)G_y = -\gamma B G_x \tag{1-33}$$

$$(d/dt)G_z = 0 \tag{1-34}$$

Interpretation of Eq. (1-34) is trivial: the component of $\mathbf{G}$ along the direction of $\mathbf{B}$ is constant. Since it has already been seen that the absolute value of **G** is unaffected, its direction and the direction of **B** must determine a time-independent angle α. The solution of the other two equations is easily obtained. If Eq. (1-32) is again differentiated with respect to time and $(d/dt)\,G_y$ is replaced by its expression given in Eq. (1-33), we arrive at

$$(d^2/dt^2)\,G_x = -(\gamma B)^2 G_x \qquad (1\text{-}35)$$

Analogous handling of Eq. (1-33) leads to

$$(d^2/dt^2)\,G_y = -(\gamma B)^2 G_y \qquad (1\text{-}36)$$

Eqs. (1-35) and (1-36) have the well-known form $(d^2/dt^2)x = -\omega^2 x$ of the harmonic oscillatory motion. The solution of Eqs. (1-32) to (1-34) is, then,

$$G_x = G \sin \alpha \cos (\omega_L t + \phi_o) \qquad (1\text{-}37)$$

$$G_y = G \sin \alpha \sin (\omega_L t + \phi_o) \qquad (1\text{-}38)$$

$$G_z = G \cos \alpha \qquad (1\text{-}39)$$

where $G \sin \alpha$ is the amplitude of the oscillatory motion and

$$\omega_L = -\gamma B \qquad (1\text{-}40)$$

is the angular velocity. The phase constant ϕ_o depends on the arbitrary choice of the origin of time. The motion is then described as a *precession* of **m** (or **G**) at constant angle α about **B**, and angular velocity $\omega_L = -\gamma B$ (Fig. 1.2). The reader should notice the analogy of these equations with the precession equations of a gyroscope spinning about an axis making an angle α with the direction of the vertical (gravity acceleration). In the next chapter, we will see that the angle α, which in the classical case depends on the initial conditions prevailing when the magnetic field was switched on, is fixed by the laws of quantization.

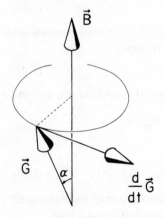

FIG. 1.2   Precession of the angular momentum of an electric charge about a magnetic field.

1.6 Magnetic field of a dipole

The magnetic field due to a dipole of magnetic moment **m** at the vector distance **r** is

$$\mathbf{B} = 3(\mathbf{m} \cdot \mathbf{r})\mathbf{r}/|r|^5 - \mathbf{m}/|r|^3 \qquad (1\text{-}41)$$

The magnetic field **B** of Eq. (1-41) may be resolved in two components: radial and angular. In order to find the expression of the radial component we perform the scalar product of **B** by the unit vector in the direction of **r**, i.e., **r**/r, as follows:

$$\mathbf{B}_r = \mathbf{B} \cdot \mathbf{r}/r = 3(\mathbf{m} \cdot \mathbf{r})\mathbf{r} \cdot \mathbf{r}/|r|^6 - \mathbf{m} \cdot \mathbf{r}/|r|^4 \qquad (1\text{-}42)$$

Since the product **m** · **r** is

$$\mathbf{m} \cdot \mathbf{r} = mr \cos \theta \qquad (1\text{-}43)$$

where θ is the angle between **r** and **m**, Eq. (1-42) reduces to

$$B_r = 2(m/|r|^3) \cos \theta \qquad (1\text{-}44)$$

The angular component B_θ may in turn be obtained after scalar multiplication of **B** by a unit vector $\mathbf{a}_o$ perpendicular to **r**, lying in the plane that contains **r** and **m**, i.e.,

$$B_\theta = 3(\mathbf{m} \cdot \mathbf{r})\mathbf{r} \cdot \mathbf{a}_o/|r|^5 - \mathbf{m} \cdot \mathbf{a}_o/|r|^3 \qquad (1\text{-}45)$$

The first term of the right-hand side of Eq. (1-45) is zero, since $\mathbf{r} \cdot \mathbf{a}_o = 0$. The scalar product $\mathbf{m} \cdot \mathbf{a}_o$ is equal to $m \cos (\frac{1}{2}\pi + \theta) = -m \sin \theta$. Therefore

$$B_\theta = (m/|r|^3) \sin \theta \qquad (1\text{-}46)$$

The angular component given by Eq. (1-46) vanishes for $\theta = 0, \pi$. The value of the field is then

$$B_{0,\pi} = B_r = \pm 2m/|r|^3 \qquad (1\text{-}47)$$

In the equatorial plane, on the other hand, $\theta = \pi/2$; the radial component of Eq. (1-42) vanishes, and the magnetic field becomes

$$B_{\pi/2} = B_\theta = m/|r|^3 \qquad (1\text{-}48)$$

The energy of interaction between two dipoles of moments $\mathbf{m}_A$ and $\mathbf{m}_B$ is given by

$$W_{d-d} = -\mathbf{m}_A \cdot \mathbf{B}_B = -\mathbf{m}_B \cdot \mathbf{B}_A \qquad (1\text{-}49)$$

Hence, using Eq. (1-41) one arrives at

$$W_{d-d} = \mathbf{m}_A \cdot \mathbf{m}_B/|r|^3 - 3\mathbf{m}_A \cdot \mathbf{r}(\mathbf{m}_B \cdot \mathbf{r})/|r|^5 \qquad (1\text{-}50)$$

Since the scalar product is commutative, $\mathbf{m}_A$ and $\mathbf{m}_B$ may be exchanged in Eq. (1-50) without affecting the result, as anticipated in Eq. (1-49).

If the dipoles are parallel, there is a particular orientation that corresponds to zero magnetic energy. In order to find the orientation at which there is no interaction, Eq. (1-50) may be set equal to zero after expressing the scalar products as trigonometric functions, i.e.,

$$m_A m_B / |r|^3 - (3 m_A m_B / |r|^3) \cos^2 \theta_0 = 0 \qquad (1\text{-}51)$$

which can only be true if

$$3 \cos^2 \theta_0 - 1 = 0 \qquad (1\text{-}52)$$

Solving Eq. (1-52) for θ_0, one arrives at

$$\theta_0 = \arccos(\sqrt{3}/3) = 54°\,44' \qquad (1\text{-}53)$$

Clearly, at this orientation the magnetic field is perpendicular to each dipole. An alternative way to find the value of θ_0 at which $\mathbf{B}$ is perpendicular to $\mathbf{m}$ is to solve Eq. (1-41) for θ_0 after setting $\mathbf{m} \cdot \mathbf{B} = 0$. The equations derived in this section are important in the study of hyperfine structure in paramagnetic resonance experiments.

1.7   The use of B and H

A word of caution should be added at this point. Usage has imposed $\mathbf{H}$ instead of $\mathbf{B}$ so that the energy of a dipole is given in the specialized literature as

$$W = -\mathbf{m} \cdot \mathbf{H} \qquad (1\text{-}54)$$

Although this may be a matter of concern if one wishes to treat para and diamagnetism rigorously, the relative difference between $\mathbf{H}$ and $\mathbf{B}$ is of order 10^{-6}-10^{-7}, so that the use of $\mathbf{H}$ instead of $\mathbf{B}$ is of no practical consequence as long as the volume susceptibility is a number very small compared to one. The tradition is so well established that one has eventually to bow to it. Accordingly, we use $\mathbf{H}$ instead of $\mathbf{B}$ in the following chapters. In the absence of magnetic materials, $\mathbf{B} = \mathbf{H}$ and are measured in the same units.

Eq. (1-54) implies that the torque given by Eq. (1-5) be rewritten as

$$\mathbf{T} = \mathbf{m} \times \mathbf{H} \qquad (1\text{-}55)$$

General References

Bleaney, B. I., and Bleaney, B., "Electricity and Magnetism," New York, Oxford University Press, 1957.

Jackson, J. D., "Classical Electrodynamics," New York, John Wiley & Sons, Inc., 1962.

Morrish, A. H., "The Physical Principles of Magnetism," New York, John Wiley & Sons, Inc., 1965.

Purcell, E. M., "Electricity and Magnetism," New York, McGraw-Hill Book Company, Inc., 1965.

2 Vector model of atoms and molecules

2.1 Quantum numbers

The magnetic properties of atoms and molecules can be completely understood only when one has a solid quantum mechanical background. Treating the problem at such a level, however, would be far beyond the scope
of an introductory book. Nevertheless, a fairly good picture of the subject
may be gained by studying, at a more elementary level, those characteristics of atoms and molecules accounting for their magnetic behavior. The
outline that follows is intended to refresh this knowledge, in the assumption that the reader has already been exposed to atomic theory. However, it is not necessarily conversant with rigorous quantum mechanical
treatments.

The state of an electron in an atom is fully described by four quantum
numbers: n, l, m_l, and m_s, whose significance follows.

(a) The principal quantum number n, whose value may be 1, 2, 3, ..., is
associated with the energy of the electron. It coincides with the principal

quantum number of Bohr's theory, related to the electron energy W_n by

$$W_n = - \frac{hRZ^2}{n^2} \tag{2-1}$$

where h is Planck's constant, Z is the atomic number, and

$$R = 2\pi^2 m_o e^2 / h^3 \tag{2-2}$$

is a universal constant known as Rydberg's number. In Eq. (2-2), m_o and e are, respectively, the mass at rest and the electric charge of the electron.

The reader will recognize in Eq. (2-1) the energy of the limit term of frequency $\nu_{n,\infty}$ belonging to the spectral series of frequencies

$$\nu_{n,m} = W_{n,m}/h = RZ^2(1/n^2 - 1/m^2) \tag{2-3}$$

for $m = \infty$, where $\nu_{n,m}$ is the frequency of the photon emitted during the transition of an electron from the level of principal quantum number $m = n + 1, n + 2, \ldots$ to the level of principal quantum number n. The corresponding hydrogen spectral series—whose transitions are shown in Fig. 2-1—are Lyman ($n = 1$), Balmer ($n = 2$), Ritz-Paschen ($n = 3$), Brackett ($n = 4$), and Pfund ($n = 5$), honoring their discoverers.

Quantum mechanical treatments confirm eq. (2-1) in the case of a one-electron atom, but this result does not hold for atoms with several electrons, owing to the electrostatic interaction among these latter. For most atoms, however, it remains true that electrons with the lower values of n have the lower energy. But in atoms having electrons which occupy, in the ground state, levels with $n = 3$ or larger, such correspondence is no longer general, for there are electron states of principal quantum number n of higher energy than states of $(n + 1)$. This situation is analyzed in Section 2.4.

The number n is directly related to the X-rays emission levels $K(n = 1)$, $L(n = 2)$, $M(n = 3)$, $N(n = 4)$, ... of heavier atoms.

Due to Pauli's exclusion principle, there can be a maximum of $2n^2$ electrons in each level (i.e., two electrons with $n = 1$, eight with $n = 2$, eighteen with $n = 3$, and so on.

(b) The orbital quantum number l may have the values $0, 1, 2, \ldots,$ $(n - 1)$ for an electron of principal quantum number n. It is related to the orbital angular momentum of the electron around the nucleus, which is

$$\hbar[l(l + 1)]^{1/2} \tag{2-4}$$

where $\hbar = h/2\pi$. Since the electron is charged, its orbital angular momentum gives rise to a magnetic moment (see Section 1.4)

$$\mathbf{m} = (-e/2m_o)\hbar[l(l + 1)]^{1/2} = -\mu_B[l(l + 1)]^{1/2} \tag{2-5}$$

where

$$\mu_B = e\hbar/2m_0 \qquad (2\text{-}6)$$

is known as the Bohr magneton.

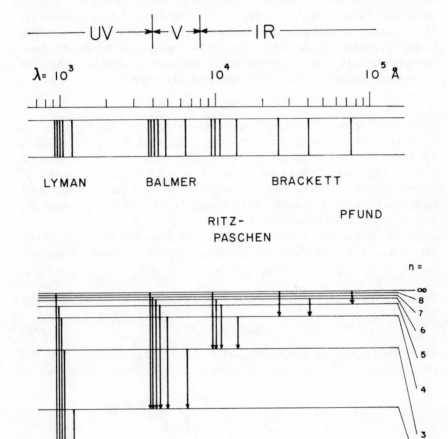

FIG. 2.1 Hydrogen spectral series.

(c) The magnetic quantum number m_l may have the values $-l$, $-(l-1), \ldots, 0, \ldots, (l-1), l$, for an electron of orbital quantum number l, and appears as a consequence of the quantized precession of the angular momentum about an applied magnetic field. As it was seen in Section 1.5, the precession theorem shows that in the presence of a magnetic field the dipole moment (and hence the angular momentum vector) will precess about the magnetic lines of force. Since precession occurs at constant angle (between the angular momentum and the magnetic field vectors), to each precession angle there corresponds a component of the magnetic moment along **H**. Quantum mechanics imposes the condition that this component shall be an integral multiple of $\hbar$, being then written

$$\hbar m_l \tag{2-7}$$

where m_l is precisely the magnetic quantum number. Notice that Eq. (2-7) is not written $\hbar[m_l(m_l + 1)]^{1/2}$ while Eq. (2-4) is written $\hbar[l(l + 1)]^{1/2}$. The meaning is that the quantized components of the angular momentum $\hbar[l(l + 1)]^{1/2}$ are proportional to m_l. This state of affairs is a general feature of quantum mechanics and will again be found in other quantum numbers (such as s, L, j, and J, to be defined later). Fig. 2-2 illustrates the polarization in a magnetic field of an electron of $l = 2$.

(d) The spin quantum number m_s may have the values $m_s = \pm\frac{1}{2}$. They determine the two quantized components of the spin angular momentum

$$\hbar[s(s + 1)]^{1/2} \tag{2-8}$$

To this angular momentum there corresponds an associated magnetic moment

$$-g_s\mu_B[s(s + 1)]^{1/2} \tag{2-9}$$

where the coefficient g_s is equal to 2.0023. The small departure from 2, the classical value, is due to a relativistic correction. The negative sign of

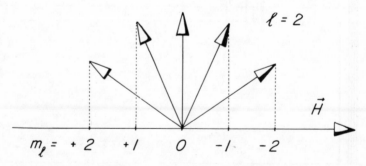

FIG. 2.2   **Spatial quantization of the orbital angular momentum in a magnetic field.**

equation (2-9) indicates that the magnetic moment of the electron is opposed to the angular momentum, as in Eq. (2-5). This fact is due to the negative charge of the electron. The two quantized components of the angular momentum of Eq. (2-8) are given by

$$\hbar m_s \tag{2-10}$$

with $m_s = \pm\frac{1}{2}$, and the corresponding components of the magnetic moment are

$$\pm \frac{1}{2} g_s \mu_B \approx \pm \mu_B \tag{2-11}$$

Two points deserve comment: The Bohr magneton is the *observed* magnetic moment of the electron due to the spin, for only the component along the magnetic field is measured, and the quality of the spin of being a half integer is compensated by the value of g_s, very close to two. The *real* spin magnetic moment of the electron would otherwise be

$$\mu_B [s(s + 1)]^{1/2} = \frac{\sqrt{3}}{2} \mu_B \tag{2-12}$$

An important principle that will often be used (and to which reference has already been made) is Pauli's exclusion principle. It states that *no atom can exist in a state in which more than one electron has the same set of four quantum numbers.* It is left to the reader to prove, on the basis of Pauli's statement, that the maximum number of electrons of principal quantum number n is $2n^2$.

2.2 Filled and unfilled shells

The term *shell* is reserved for the set of all possible states of a certain value of the principal quantum number n. Since such a shell will in general have $2n^2$ available states, if these states are all occupied by electrons it is said that the shell is *filled*; otherwise, it is referred as an *unfilled* shell. We then extend the concept to a certain value of l and we have, analogously, filled and unfilled *subshells*. The K-shell ($n = 1$) has one trivial subshell ($l = 0$). The L-shell ($n = 2$) has two subshells: $l = 0, 1$. In general, the shell of principal quantum number n has n subshells. Electrons belonging to subshells of $l = 0, 1, 2, 3$, are respectively called s-, p-, d-, f-electrons. This terminology has been borrowed from the classical labeling of spectral series, namely: sharp (s), principal (p), diffuse (d), and fundamental (f). Further labeling of electrons with $l = 4, 5, \ldots$ has simply been made by using the a.phabetic order $g, h, \ldots$. Occurrence of filled shells in different elements gives rise to the well-known periodic similarity of properties with respect to the atomic number Z. From the point of view of the magnetic behavior, the most important property of a filled shell or subshell is that the resultant magnetic moment of the electrons is zero. This property can be proved by using the following arguments. If the shell is filled, there will be two electrons for each set of equal n, l, and m_l; one of them having $m_s = +\frac{1}{2}$, the other: $m_s = -\frac{1}{2}$. The spin magnetic moments must then cancel out. Analogously, electrons having $m_s = +\frac{1}{2}$, will have $m_l = 0$, which does not count from the magnetic point of view. In addition, for each $m_l \neq 0$ there will be an $m_l \neq 0$ of equal absolute value but opposite sign (i.e., $m_l = \pm l$, $\pm(l - 1)$, etc.). The orbital angular magnetic moments will cancel out for each value of m_s and hence the whole system of electrons in a filled shell or subshell cannot have a resultant magnetic moment. An important consequence of this argument is that in order for an atom to have a magnetic moment, it must have partially filled subshells. But for evaluating the resultant magnetic moment, it is necessary to find the coupling mechanism among the different contributions to the total magnetic moment (i.e., the way in which orbital and spin magnetic moments couple among themselves).

2.3   Coupling of angular and spin momenta

The coupling mechanism of angular and spin momenta among different electrons in an atom clearly depends on the mutual interaction among them, due to their electric charge. When this interaction is treated by quantum mechanical theoretical procedures, an unexpected result is found, which has no analog in classical theory: the existence of an *exchange energy* that, by overcoming the electrostatic repulsion, leads to a strong coupling between electron spins. This coupling is so strong that, within an atom, the state with the spins parallel may have the lower energy and hence be more stable. This interaction energy is usually written

$$W_e = -2J\mathbf{s}_i \cdot \mathbf{s}_j \tag{2-13}$$

where $\mathbf{s}_i$, $\mathbf{s}_j$ are the spin vectors, and J is a scalar quantity called exchange energy.* Thus, with a number of electrons the exchange forces produce the coupling of the spin vectors $\mathbf{s}_i$, $\mathbf{s}_j$ that leads to the maximum value

---

*Unfortunately, the number of available symbols is rather limited. J is therefore used for the exchange energy as well as for the total angular momentum of an atom. The reader must be careful not to interchange meanings.

$S = \sum_k s_k$ permitted by Pauli's exclusion principle. A similar phenomenon occurs with the orbital vectors l that couple to give a maximum value $L = \sum_k l_k$, also consistent with Pauli's principle. That both S and L must have maximum values is known as one of Hund's rules. Curiously enough, lower values of S and L are possible, for they are consistent with Pauli's principle, but correspond to states of higher energy. This coupling mechanism is known as Russell-Saunders coupling.

Both S and L have absolute magnitudes of the total spin and orbital angular momenta

$$\hbar[S(S + 1)]^{1/2} \tag{2-14}$$

$$\hbar[L(L + 1)]^{1/2} \tag{2-15}$$

Both S and L are *good* quantum numbers as long as coupling is not destroyed, i.e., they behave as the spin and orbital angular momenta in an atom with only one unpaired electron in an unfilled subshell.

An alternate, less frequently encountered way of coupling is the so-called *j-j* coupling. In it, the vectors s and l of each electron couple together to give a resultant j. Then the different j's add to give the total angular momentum J. In general, the s_p-s_q interaction is stronger than the s_p-l_p interaction, favoring Russell-Saunders coupling, except for very heavy elements. Subscripts are used to label electrons.

The Russell-Saunders vectors S and L interact with an energy $\lambda L \cdot S$, where λ is called the *spin-orbit coupling constant,* to give the total angular momentum

$$J = L + S \tag{2-16}$$

whose absolute value is

$$\hbar[J(J + 1)]^{1/2} \tag{2-17}$$

As long as coupling between S and L is not destroyed, J is also a good quantum number. It may be integer or half-integer, since S may be integer or half-integer.

2.4 Nomenclature of electronic configurations and atomic energy states

As we have seen in Section 2.2, electrons with $l = 0, 1, 2, 3, 4, \ldots$ are called *s-, p-, d-, f-, g-, \ldots* electrons. It is customary to precede each letter with the numerical value of the principal quantum number n. Thus, we have $1s$, $2s$, $2p$, $3s$, $3p$, $3d$, $4s$, ... electrons. The number of electrons in each state (subshell) is indicated by a right superscript ($1s^2$, $3p^6$, $4d^7$, ...). Since for each value of n there can be no more than two s-electrons, six p-electrons, ten d-electrons, fourteen f-electrons, etc., filled shells are $1s^2$, $3p^6$, $4d^{10}$, $4f^{14}$, while $2s$, $3p^5$, $3d^8$, $4f^9$ represent unfilled or, less ambiguously, partially filled subshells.

It is then clear that the ground state of the hydrogen atom is $1s$, since it has only one electron with $n = 1$. Thus, the helium electronic configuration will be ($1s^2$), that of lithium: ($1s^2 2s$). After Be($1s^2 2s^2$), electrons begin to fill the $2p$-shell. Neon, the second rare gas, has the configuration ($1s^2 2s^2 2p^6$). The process is continued until the third rare gas Ar($1s^2 2s^2 2p^6 3s^2 3p^6$). After argon, it no longer holds true that the lower energy corresponds to the lower value of n. In potassium, which immediately follows argon in atomic number, the new electron does not occupy the $3d$ but the $4s$ level. The electronic configuration of potassium may then be written (Ar, $4s^1$), representing the configuration of the core by the symbol of the rare gas. It is after Ca(Ar, $4s^2$) that the $3d$ level begins to be filled, giving rise to the first transition series, gathering elements of electronic configuration (Ar, $4s^2 3d^r$) with $1 \leq r \leq 10$. Electrons then begin to fill the $4p$ level until the third rare gas: Kr(Ar, $4s^2 3d^{10} 4p^6$). The sequence is repeated for the $4d$ levels after filling the $5s$ levels (Rb and Sr), giving rise to the second transition series that ends in Xe(Kr, $5s^2 4d^{10} 5p^6$). Next come Cs(Kr, $6s^1$) and Ba(Kr, $6s^2$), and then the $4f$-subshell becomes the lowest-energy level.

Elements with unfilled $4f$ levels are the rare earths (or lanthanides). Once the $4f$-subshell is filled, electrons start to occupy the $5d$-subshell, giving rise to the third transition series.

Finally, those with unfilled $5f$-subshells are the actinides, including the transuranian elements with two electrons in $7s$, and sometimes one in $6d$. Although there are exceptions, a mnemonic rule, shown in Fig. 2.3, gives the filling order as a path through the subshell array, starting in $1s$. Appendix I (p. 208) lists the electronic configurations of free atoms.

The notation developed in preceding paragraphs, though physically meaningful, is still complex for handling. A reduced notation has been developed in which a capital letter with a left superscript and a right subscript describes the spectroscopic and magnetic characteristics of the free atom (or ion). The capital letter represents the resultant orbital angular momentum of the system and has been chosen by capitalizing the corresponding letters used for designating electrons on the basis of their orbital angular momentum. Thus, the capital letters S, P, D, F, G, etc., indicate a resultant orbital angular momentum $L = 0, 1, 2, 3, 4$, etc. The left superscript is equal to $2S + 1$, where S stands for the resultant spin.

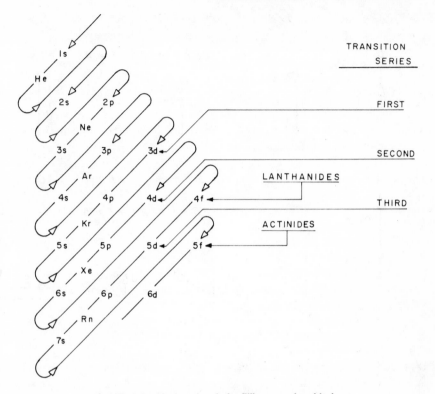

FIG. 2.3 Mnemonic rule for filling atomic orbitals.

Notice that in order to avoid misunderstandings, italic S and bold-face roman **S** are used for spin and spin vector, while roman capital S is reserved to designate the state. The right subscript is equal to $J = L + S$, where, according to the second Hund's rule, addition must be performed if the subshell is more than half occupied, and subtraction if less than half occupied. When the subshell is half occupied, $L = 0$, and then $J = S$.

Atoms with all filled subshells have $S = L = 0$; their ground state (term) is ${}^1S_0$. Cupric ion $Cu^{2+}(Ar, 3d^9)$, for example, must be a ${}^2D_{5/2}$ ion, since $S = \frac{1}{2}$, $L = 2$, and L and S must be added, for the subshell is more than half occupied. Atomic luthecium, on the other hand, having only one unpaired d-electron, is described by the term ${}^2D_{3/2}$.

The meaning of the left superscript $2S + 1$ is the actual multiplicity of the energy levels when $L \geq S$, though not when $L < S$, for in the former case there are $2S + 1$ ways to combine S and L, while there are only $2L + 1$ ways in the latter one. This situation is exemplified in Fig. 2.4 where the cases $(S = 3/2, L = 2)$, $(S = L = 2)$, and $(S = 2, L = 1)$

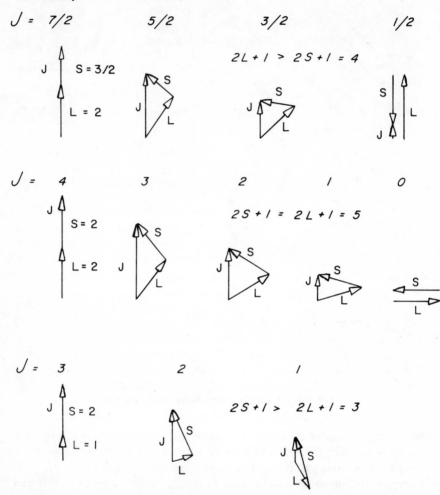

FIG. 2.4   **Different cases of spin-orbit coupling.**

are shown. The multiplicity of the term is actually given by either $2S + 1$ or $2L + 1$, whichever is less.

2.5   Magnetic moments of free atoms

It was seen in Section 2.3 that the total spin and total orbital momentum are given by

$$\hbar[S(S + 1)]^{1/2} \tag{2-14}$$

and

$$\hbar[L(L + 1)]^{1/2} \tag{2-15}$$

while the total momentum represented by the vector $\mathbf{J} = \mathbf{S} + \mathbf{L}$ of Eq. (2-16) is

$$\hbar[J(J + 1)]^{1/2} \tag{2-17}$$

The magnetic moments associated with the $\mathbf{S}$ and $\mathbf{L}$ vectors are, respectively:

$$m_{\mathrm{L}} = -\mu_{\mathrm{B}}[L(L + 1)]^{1/2} \tag{2-18}$$

$$m_{\mathrm{S}} = -g\mu_{\mathrm{B}}[S(S + 1)]^{1/2} \tag{2-19}$$

using analogous arguments to those used in the derivation of Eqs. (2-5) and (2-9).

The problem now is to find the magnetic moment associated with $\mathbf{J}$. Since each vector $\mathbf{S}$ and $\mathbf{L}$ involves a magnetic moment, if the $\mathbf{S}$ and $\mathbf{L}$ vectors are coupled, each one must precess about the magnetic field due to the other, according to the precession theorem studied in Section 1.6. In the absence of external torques acting on the system, the total angular momentum $\mathbf{J}$ will be constant and fixed in space. As a consequence, $\mathbf{L}$ and $\mathbf{S}$ must precess about $\mathbf{J}$. Since the orbital and spin magnetic moments are parallel to $\mathbf{L}$ and $\mathbf{S}$, they, as well as their resultant $\mathbf{m}$, must also precess about $\mathbf{J}$. Notice that $\mathbf{m}$ is no longer parallel to $\mathbf{J}$, due to the spin g-value, equal to two. Fig. 2.5 shows the multiple vector diagram relating mag-

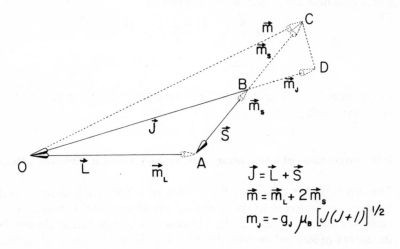

FIG. 2.5   Association of magnetic moments in spin-orbit coupling.

netic moments and angular momenta, respectively indicated in dotted and solid lines. The length of the $\mathbf{m}_L$ vector has arbitrarily been set equal to the length of the $\mathbf{L}$ vector. Consequently, the $\mathbf{m}_S$ vector is twice the $\mathbf{S}$ vector, due to the value of g_S, approximately two. From the geometry of the triangle AOB one obtains

$$-\cos(AOB) = \frac{S(S + 1) - L(L + 1) - J(J + 1)}{2[L(L + 1)J(J + 1)]^{1/2}} \tag{2-20}$$

and the projection of the orbital momentum on $\mathbf{J}$ is then

$$\mu_B \frac{S(S + 1) - L(L + 1) - J(J + 1)}{2[J(J + 1)]^{1/2}} \tag{2-21}$$

The projection of the spin moment on $\mathbf{J}$ is given by

$$m_S \cos(OBA) = -g_S\mu_B[S(S + 1)]^{1/2}\cos(OBA)$$

$$= 2\mu_B \frac{L(L + 1) - S(S + 1) - J(J + 1)}{2[J(J + 1)]^{1/2}} \tag{2-22}$$

The magnetic moment m_J results from adding Eq. (2-21) and (2-22):

$$m_J = \mu_B \frac{L(L + 1) - S(S + 1) - 3J(J + 1)}{2[J(J + 1)]^{1/2}} \tag{2-23}$$

In order to have an expression formally consistent with Eqs. (2-18) and (2-19), we may write

$$m_J = -g_J\mu_B[J(J + 1)]^{1/2} \tag{2-24}$$

where g_J is now the Landé factor, defined by

$$g_J = \frac{3}{2} + \frac{S(S + 1) - L(L + 1)}{2J(J + 1)}$$

$$= 1 + \frac{J(J + 1) + S(S + 1) - L(L + 1)}{2J(J + 1)} \tag{2-25}$$

In the special case of spin-only paramagnetism, Eq. (2-25) leads to the free electron value $g_S = 2$ since $L = 0$.

2.6   Interaction of a free atom with a magnetic field: Zeeman effect

The situation that arises when an atom with angular and spin magnetic moments is subjected to the action of a magnetic field is complex. Two extreme cases, however, can be handled with reasonable accuracy: when the energy of interaction with the magnetic field is either much smaller or much larger than the *L-S* coupling energy. In the first case, the *L-S* coupling is not affected and the magnetic vector associated with J quantizes in the direction of the field; its components have the values $m_J = -J, -(J - 1), \ldots, (J - 1), J$. The component of the magnetic moment of the atom, parallel to the magnetic field, is then

$$-m_J g_J \mu_B \qquad (2\text{-}26)$$

The degeneracy (equal energy) of all m_J levels of a certain value of j is then lifted by the magnetic field. This fact, together with the existence of states with other values of j, gives rise to the Zeeman effect. Thus, in a singlet state for which $J = 1$ ($S = 0$), and hence $m_J = -1, 0, 1$, the frequency difference between two consecutive emission lines, which corresponds to $\Delta m_J = \pm 1$, is

$$\Delta \nu = \mu_B H / h \qquad (2\text{-}27)$$

since $g(J = 1) = 1$. The difference $\Delta \nu$ refers to the frequency difference (in second^{-1}) in the emission lines resulting from bringing an electron from infinity to occupy each Zeeman level characterized by a value of m_J. The three emission lines correspond to the three values of m_J; they are called the Lorentz triplet of a singlet state. Since, in most cases, the L-S coupling energy is equivalent to the coupling energy of a free electron with a field of the order of 10^5 oersted, this approximation is valid for ordinary magnetic fields available in a laboratory.

In the case of light elements, in which the L-S coupling energy is significantly smaller, and when the magnetic field is exceptionally large,* the Russell-Saunders coupling is broken down and the **S** and **L** vectors independently precess about the magnetic field **H**. In the such a case, **J** is not a *good* quantum number and the inadequacy of the former approximation may be verified when large fields are applied to light atoms. One then observes a departure from the linear increase in the separation of the predicted Zeeman components with the value of H. This is known as the Paschen-Back effect, named after its discoverers.

2.7 Degeneracy of energy levels

An important feature of electron energy levels in a free atom is that electrons in the same subshell all have the same energy, although, according to Pauli's principle, they differ in at least one quantum number. Such levels are called *degenerate*. Degeneracy may be lifted by electric and mag-

---

*Very large pulsed magnetic fields may be produced by periodic discharges; very large steady magnetic fields may be obtained with superconducting coils.

netic fields, and by the chemical environment. This chemical environment is reduced to a purely electrostatic field in the crystal field theory. Removal of degeneracy is governed by the laws of symmetry. Although this subject is treated in detail in Chapters 5, 7, and 12, this section will briefly state the reasons that account for such a behavior. Let us consider, for example, the six electrons of a p-subshell. Solution of Schrödinger's equation leads to the assignment of three independent angular functions to the p-levels, which may be chosen as proportional to x, y, and z, respectively (see Section 12.2). The p-levels may then be labeled p_x, p_y, and p_z, each of them accepting two electrons of opposite spin. In an environment of cubic symmetry, where x, y, and z are naturally chosen as parallel to the three edges of a cube, x, y, and z are equivalent, for the operations of symmetry that characterize such a configuration exchange x, y, and z. For example, a rotation by $\frac{1}{2}\pi$ about z exchanges x and y; the same rotation about y exchanges z and x. Analogously, a rotation by $2\pi/3$ about a diagonal of the cube exchanges x, y, and z. All these operations will analogously exchange the orbitals p_x, p_y, and p_z. However, since they are operations of symmetry, p_x, p_y, and p_z must be equivalent. Thus, the p-level is degenerate. There are then, six p-states that have the same energy in a cubic environment. Since there is spherical symmetry in a free atom —which also exchanges x, y, and z—the p-level is also degenerate. The six p-states may be gathered in three sets differing in magnetic quantum number ($m_l = -1, 0, 1$), each set having two electrons which, in turn, differ in spin quantum number ($m_s = \pm\frac{1}{2}$). It can be said that the total degeneracy of the p-level is six, and that it has three-fold orbital degeneracy. This leaves a two-fold spin degeneracy in each orbital p_x, p_y, p_z. If we now distort the cube along the z-axis, the new environment will have tetragonal symmetry. Its symmetry operations will now exchange x and y, but not z. Then p_x and p_y will still have the same energy, but not p_z. The new environment has then partially removed the orbital degeneracy and the new scheme of energy levels will have two p-states (p_x and p_y) with two-fold orbital degeneracy, and one p-state (p_z) having only spin degeneracy. Further distortion of the environment will eventually remove the equivalence between x and y and completely lift the orbital degeneracy. An important theorem on this subject is *Kramer's theorem* which states that a purely electrostatic field acting upon a system of an odd number of electrons can never reduce its total degeneracy below two, for no electrostatic field can remove the spin degeneracy. A magnetic field is necessary to remove the spin degeneracy, and this effect is precisely the cause of paramagnetism. Another important theorem is due to Jahn and Teller. It states that a nonlinear molecule having an orbitally degenerate ground state will spontaneously distort itself so as to remove the orbital degeneracy. This is the reason that explains why, for example, octahedral environments of water molecules around a hydrated ion, of the type $Me(H_2O)_6^{n+}$, almost invariably are slightly distorted (even in cubic crystals) and have, in general, orthorhombic symmetry. It will be seen later that the removal of orbital degeneracy by the chemical environment accounts, among other facts, for optical absorption arising from transitions between levels split by the ligand field.

2.8 Quenching of angular momenta

Since removal of degeneracy is determined by electrostatic interaction between the electron angular wave function and the environment, it is necessary to determine the energy involved in this interaction. It is found that when the unpaired electron is an outer electron, therefore being unscreened by other electrons of the same atom, it strongly interacts with the environment, to the extent of breaking down the L-S coupling. In this case, it is observed that unpaired electrons having large angular momenta behave in the presence of a magnetic field as if they had negligible angular momentum. Their behavior approaches that of spin-only paramagnetism, i.e., their g-value is close to two despite their having magnetic quantum numbers other than zero. We then say that the angular momentum is *quenched*. Quenching may, of course, be partial, and practically any value of g between zero and the free-atom one may be observed. This effect is particularly remarkable in organic radicals and ions of the transition series (d-ions). In the ions having unpaired f-electrons, lanthanides and actinides, these electrons are screened by the ion outer electrons and their interaction with the ligand field is often weak. Due to these circumstances, the amount of quenching is correspondingly small, and the g-values approach the free-ion value.

2.9 Vector model of free molecules

Molecules with three or more atoms have directed valence bonds and need to be treated by hybridizing atomic orbitals. This subject is beyond the scope of this book. Diatomic molecules may be studied by the *united-atom* approach. In the united-atom approach, the atomic orbitals have resolved angular momenta along the symmetry axis of the molecules. Orbitals with resolved angular momentum $\lambda = 0, 1, 2, 3$, etc., are labeled $\sigma, \pi, \delta, \phi$, etc., which are the molecular counterparts of the atomic symbols $s, p, d, f, \ldots$. Different orbitals $ns, np, \ldots$ are then symbolized according to Table 2.1.

TABLE 2.1 United-atom wave functions, resolved angular momenta $(m\lambda)$, and state symbols

Wave Function	m_λ	Symbol for State
ns	0	$ns\,\sigma$
np_0	0	$np\,\sigma$
$np_{\pm 1}$	± 1	$np\,\pi$
nd_0	0	$nd\,\sigma$
$nd_{\pm 1}$	± 1	$nd\,\pi$
$nd_{\pm 2}$	± 2	$nd\,\delta$

The total resolved angular momentum along the axis

$$\Lambda = \Sigma \lambda_i \tag{2-29}$$

takes the place of the atomic expression $L = \Sigma l_i$. To the atomic $S = \Sigma s_i$ there corresponds the molecular Σ, and to the atomic $J = L + S$ there corresponds

$$\Omega = \Lambda + \Sigma \tag{2-30}$$

The terminology for the state of a molecule is then established as follows: The Greek capitals Σ, Π, Δ, Φ,...(borrowed from the atomic terminology S, P, D, F,...) indicate $\Lambda = 0, 1, 2, 3,\ldots$. Care should be taken to avoid confusion between the symbol Σ for the total spin and the same symbol representing states of $\Lambda = 0$. The left superscript is $2\Sigma + 1$, and the right subscript is Ω.

In this way, the ground state of the free NO molecule, for example, of electronic configuration

$$NO = (1s\sigma)^2(2p\sigma)^2(2s\sigma)^2(3p\sigma)^2(3d\sigma)^2(2p\pi)^4(3d\pi)$$

is written $^2\Pi_{1/2}$, with the meaning of $\Lambda = 1$, $\Sigma = \frac{1}{2}$, and $\Omega = \frac{1}{2}$. The NO molecule is of unusual theoretical importance because it is nonparamagnetic in the $^2\Pi_{1/2}$, despite the existence of one unpaired electron.

General References

Bates, L. F., "Modern Magnetism," Fourth Ed., Cambridge, England, Cambridge University Press, 1961.

Bleaney, B. I., and Bleaney, B., "Electricity and Magnetism," New York, Oxford University Press, 1957.

Herzberg, G., "Atomic Spectra and Atomic Structure," New York, Dover Press, 1944.

Kauzmann, W., "Quantum Chemistry," New York, Academic Press, 1957.

3 Theory of magnetic susceptibilities

3.1 The magnetic properties of matter

Substances with a negative magnetic susceptibility are called *diamagnetic*. Substances with a positive magnetic susceptibility, independent of the applied magnetic field,* are called *paramagnetic*. While diamagnetism is practically temperature independent, paramagnetism generally is not.†

Ideally, isotropic paramagnetic substances follow Curie's law

$$\chi = C/T \tag{3-1}$$

which determines an equation of state relating the magnetic field H, the magnetization of the sample M, and the temperature T:

$$MT = CH \tag{3-2}$$

where C is a constant.

* The susceptibility is not a function of the magnetic field for moderate values of the latter and not very low temperatures. Otherwise, saturation occurs and the magnetic susceptibility approaches a maximum value. See Section 3.3.

† Paramagnetism of conductivity electrons in metals is temperature independent. See Section 3.5.

Out of the scope of this book are other forms of magnetization: ferro- and ferrimagnetism. Their phenomenological characteristics should, however, be known, for such behaviors often appear in paramagnetic substances at low temperature.

Ferromagnetism occurs below the so-called temperature of the Curie point, T_C. The magnetic susceptibility increases tremendously depending on the magnetic field. In addition, ferromagnetic substances become permanently magnetized and exhibit saturation at fields of a few kilo-oersted. Above T_C, the susceptibility no longer depends on the magnetic field, and follows the Curie-Weiss law

$$\chi = \frac{C}{T - T_C} \tag{3-3}$$

Antiferromagnetism is characterized by a well-defined kink in the susceptibility-*vs.*-temperature curve. Below the temperature of the transition, the magnetic susceptibility *decreases* with decreasing temperature while, above the transition point, the magnetic susceptibility decreases with increasing temperature, approaching the ideal paramagnetic behavior.

Returning to diamagnetism and paramagnetism, it is an accepted practice in introductory books to develop the classical Langevin theory of magnetic susceptibilities. We will not abandon this practice, because the classical theory conveys a picture of both phenomena that is helpful in understanding the magnetic behavior of matter.

Lenz's law of electromagnetism states that if the magnetic flux through an electric circuit is varied, a counterelectromotive force is generated, which in turn gives rise to an electric current that creates a magnetic field opposing the change. This is the property used to define the induction vector **B** in Eq. (1-6). Though Lenz's law clearly accounts for *transient* phenomena in resistive circuits, its application to resistanceless circuits (electron orbits) must give rise to a permanent current, henceforth to a steady magnetic moment opposed to the applied magnetic field. This state of affairs will result in a field inside the system *smaller* than that outside. The existence of these induced magnetic moments well accounts for the diamagnetic behavior.

On the other hand, if there is associated to each atom, ion, or molecule a permanent magnetic moment, each atom, ion, or molecule will tend to line up its magnetic dipole parallel to the magnetic field (position of minimum potential energy). The existence of thermal agitation will tend to restore random orientation. As a consequence of these two competitive processes, there will be a macroscopic (resultant) magnetic moment that will increase with the magnetic field and decrease with increasing temperature. This behavior is quantitatively accounted for by the equation of state (3-2) of paramagnetic substances.

3.2 Diamagnetic susceptibility

The argument used in discussing the application of Lenz's law to a resistanceless circuit may be put in a simple quantitative way if the problem is restricted to monoatomic molecules, for in this case Larmor's theorem may be used.

Larmor's theorem states, briefly, that if $z_i(t)$, $r_i(t)$, and $\phi_i(t)$ are the cylindrical coordinates as a function of time of the ith. electron of an atom with the nucleus at the origin, in the absence of a magnetic field, then application of a magnetic field **H** along the z-axis only effects (to the first order in H) the angular function $\phi_i(t)$ which takes the form

$$\phi_i(t) = \phi_i^o(t) + \omega_L t \qquad (3\text{-}4)$$

where the superscript o indicates the value for $H = 0$, and

$$\omega_L = -\gamma H \qquad (3\text{-}5)$$

is the angular velocity already introduced in equation (1-40). This theorem follows immediately from the general precession theorem developed in Section 1.5. The fact that it is valid to the first order in H is a consequence of treating the specific case of an electron orbiting about a nucleus, subject to a central coulombian force. The general precession theorem developed in Section 1.5 is more general in its derivation; its initial postulate is only the connection between angular momentum and magnetic dipole moment.

Replacing γ by the expression given in Eq. (1-29) after taking into account that $q = -e$ is, in this case, the charge of the electron, the angular velocity of precession becomes

$$\omega_L = eH/2m_o \qquad (3\text{-}6)$$

or, in frequency units

$$\nu_L = (2.80 \times 10^6 \text{s}^{-1} \text{gauss}^{-1}) H \qquad (3\text{-}7)$$

Eq. (3-4) is a good approximation provided that the force exerted on the electron by the applied magnetic field is small compared with the force exerted by the nucleus, which is certainly true under ordinary conditions. The effect of the magnetic field on the electronic motion is to cause each electronic orbit to rotate about the direction of **H**, or more precisely, **B**, with the angular velocity ω_L.

As a consequence of the Larmor precession, each electron acquires a component of angular momentum along the direction of the magnetic field

$$G = m_o \langle a^2 \rangle \omega_L = m_0 \langle a^2 \rangle (e/2m_o) H \qquad (3\text{-}8)$$

where $\langle a^2 \rangle$ is its mean square distance to the axis parallel to **H** that passes through the nucleus. The expression $m_o \langle a^2 \rangle$ is the moment of inertia of the electron with respect to said axis. The mean square distance $\langle a^2 \rangle$ is the average of the perpendicular distance of the electron to the field axis. In an orthogonal frame with the origin at the nucleus, it may be represented by

$$\langle a^2 \rangle = \langle x^2 \rangle + \langle y^2 \rangle \tag{3-9}$$

after setting the z-axis parallel to **H**. In terms of the mean square distance $\langle r^2 \rangle$ of the electron to the nucleus.

$$\langle r^2 \rangle = \langle x^2 \rangle + \langle y^2 \rangle + \langle z^2 \rangle \tag{3-10}$$

we have

$$\langle a^2 \rangle = \frac{2}{3} \langle r^2 \rangle \tag{3-11}$$

On summing over all electrons in the atom we obtain

$$\Sigma \langle a^2 \rangle = \frac{2}{3} \Sigma \langle r^2 \rangle \tag{3-12}$$

To the angular momentum of Eq. (3-8) there is associated a magnetic moment

$$\mathbf{m} = -(e/2m_o)\mathbf{G} \tag{3-13}$$

given by Eqs. (1-28) and (1-29) after replacing $-e$ for q. Introducing the value of G given by Eq. (3-8) in terms of $\langle r^2 \rangle$ (Eq. (3-11)) and summing over all electrons, each atom then acquires a magnetic moment

$$m = -(e^2/6m_o)H \Sigma \langle r^2 \rangle \tag{3-14}$$

Consequently, the susceptibility of a substance containing n atoms per unit volume will be

$$\kappa = nm/H = -n(e^2/6m_o)\Sigma \langle r^2 \rangle \tag{3-15}$$

Eq. (3-15) shows that the diamagnetic susceptibility is inherently negative in sign and does not depend on the sign of the electronic charge, which was to be expected, for the negative sign follows from Lenz's law.

If one defines the average

$$\langle R^2 \rangle = (1/Z)\Sigma \langle r^2 \rangle \tag{3-16}$$

for an atom of atomic number Z, Eq. (3-15) takes the form

$$\kappa = -\frac{Ze^2 n}{6m_o} \langle R^2 \rangle \tag{3-17}$$

which is the Langevin expression as corrected by Pauli.* Quantum mechanical treatment of diamagnetism leads to the same result. The problem of calculating the diamagnetic susceptibility is thus reduced to the evaluation of $\langle R^2 \rangle$. Such an evaluation amounts to determining the electron charge distribution within the atom. Several methods have been developed to deal with the problem.[1,2,6,8] In the simple cases, the accuracy of the theoretical estimates is striking. In the case of helium, for instance, the calculated gram-atom susceptibility for an estimated average radius $\langle R^2 \rangle = 1.1848$ a. u. is $\chi = -1.878 \times 10^{-6}$ emu,[5] while the experimental value is -1.88×10^{-6} emu.[9] In the case of neon, using $\langle R^2 \rangle(1s) = 0.03347$, $\langle R^2 \rangle(2s) = 0.96967$, and $\langle R^2 \rangle(2p) = 1.2381$, the calculated value is -7.475×10^{-6} emu, while the observed one is $-(6.95 \pm 0.14) \times 10^{-6}$ emu.[7]

Since the value of $\langle R^2 \rangle$ is practically unaltered by temperature, the susceptibility of a diamagnetic substance is temperature independent to a good approximation. In addition, $\langle R^2 \rangle$ is practically a constant for each atom or ion. This explains the fact that aqueous solutions of alkali and alkaline earth halides follow quite accurately the additivity rule of Wiedemann, which states that the mass susceptibility χ_m of a solution containing a mass m_1 of salt of susceptibility χ_1 and a mass m_2 of solvent of susceptibility χ_2 is given by

$$\chi_m = \frac{m_1 \chi_1 + m_2 \chi_2}{m_1 + m_2} \tag{3-18}$$

Although this rule is only approximate, it is obeyed by many substances, expecially by compounds that ionize in solution. Myers has reviewed the diamagnetism of ions.[4] Evaluation of the diamagnetic susceptibility of organic molecules is complicated by the fact that the law of additivity must be supplemented with structural contributions, and by the frequently strong anisotropy of ring molecules. More information on this subject is given in Appendix III.

3.3 Paramagnetic susceptibility

The classical derivation of Curie's law was given by Langevin, using Boltzmann's statistics. His basic assumption was that each atom or molecule had a permanent magnetic moment **m** and that the only force acting on it was the resultant of the applied magnetic field **H**. In order to determine the orientation distribution of these magnetic dipoles as a con-

---

*Langevin's original equation did not include the factor $^2/_3$ of Eq. (3-12).

sequence of the competitive process of alignment in the magnetic field and the thermal agitation, it is necessary to start by establishing the magnetic potential energy of a magnetic dipole in a magnetic field. This is given by

$$W = -\mathbf{m} \cdot \mathbf{H} = -mH \cos \theta \tag{3-19}$$

where θ is the angle between the directions of $\mathbf{m}$ and $\mathbf{H}$. The physical interpretation of Eq. (3-19) follows. Since the scalar product $\mathbf{m} \cdot \mathbf{H}$ vanishes when $\mathbf{m}$ and $\mathbf{H}$ are mutually perpendicular, we take the magnetic energy as zero. Then, when the magnetic moment $\mathbf{m}$ lines up parallel to the magnetic field, it has negative potential energy, since in order to rotate it it would be necessary to do work. This explains the negative sign of Eq. (3-19). The work done on the magnetic dipole of magnetic moment $\mathbf{m}$, to place it perpendicular to $\mathbf{H}$, is clearly mH. If the dipole lines up antiparallel to $\mathbf{H}$, it will have rotated by π from the parallel position about an axis perpendicular to $\mathbf{H}$. The new value of the potential energy will now be positive, and expressed by the fact that $\cos \pi = -1$. The choice of zero magnetic energy when $\mathbf{m}$ and $\mathbf{H}$ are mutually perpendicular arises from the fact that the energy of the dipole, under such conditions, is not affected if the magnetic field is switched off. Eq. (3-19) results from integrating the torque $\mathbf{T}$ given by Eq. (1-55) from $\theta = \frac{1}{2}\pi$ to θ, i.e.,

$$W = \int_{1/2\pi}^{\theta} \mathbf{m} \times \mathbf{H} \cdot d\theta = mH \int_{1/2\pi}^{\theta} \sin \theta \, d\theta = -mH \cos \theta \tag{3-20}$$

and accounts for the interpretation given above.

If a paramagnetic gas is considered, in order to neglect magnetic interactions among molecules, the number of molecules having an energy in the range W to $W + dW$ is given by Boltzmann's statistics as

$$dn = c \exp(-W/kT)dW \tag{3-21}$$

where k is Boltzmann's constant, T is the absolute temperature, and c is a constant chosen in such a way that integration of Eq. (3-21) over all possible values of W shall be equal to the total number of molecules N. We may now replace W in Eq. (3-21) by its expression given in Eq. (3-19), with the result

$$dn = c \exp[(mH \cos \theta)/kT] d(-mH \cos \theta)$$
$$= c \exp[(mH \cos \theta)/kT] \cdot mH \sin \theta \, d\theta \tag{3-22}$$

The total number N of molecules must then be equal to Eq. (3-22) integrated over all angles between 0 and π.

The net magnetization of the sample is given by the resultant of all magnetic moments along the direction of **H**. The component of each dipole is

$$m \cos \theta \tag{3-23}$$

The average magnetic moment along the direction of **H** is then

$$\langle m \rangle = (1/N) \int_N m \cos \theta \, dn \tag{3-24}$$

the integral being extended over the total number N of molecules. Hence, the total magnetic moment along the direction of **H** is

$$N \langle m \rangle = m \int_N \cos \theta \, dn \tag{3-25}$$

and then

$$\frac{\langle m \rangle}{m} = \frac{\int_0^\pi \cos \theta \cdot c \exp(mH \cos \theta / kT) \cdot mH \sin \theta \, d\theta}{\int_0^\pi c \exp(mH \cos \theta / kT) \cdot mH \sin \theta \, d\theta} \tag{3-26}$$

On writing

$$a = mH/kT \tag{3-27}$$

$$x = \cos \theta \tag{3-28}$$

Eq. (3-26) takes the form

$$\frac{\langle m \rangle}{m} = \frac{\int_{-1}^{+1} x \exp(ax) \, dx}{\int_{-1}^{+1} \exp(ax) \, dx} = \coth a - 1/a = L(a) \tag{3-29}$$

where $L(a)$ is known as the Langevin function of a. The Langevin function tends to unity with increasing a. This property is consistent with saturation of paramagnetic substances, observed at low temperature for large values of the magnetic field, although it does not predict the correct value. Saturation obviously occurs when all the dipoles line up parallel to the magnetic field.

For small values of a (i.e. for magnetic potential energies mH small compared with kT) $L(a)$ is, to a good approximation, linear in a

$$\langle m \rangle / m = L(a) \approx \frac{1}{3} a = mH/3kT \tag{3-30}$$

Since the magnetic molar susceptibility is given by

$$\chi = M/H = N\langle m \rangle / H \tag{3-31}$$

replacing $\langle m \rangle$ by its value given in Eq. (3-30) leads to

$$\chi = \frac{Nm^2}{3kT} \tag{3-32}$$

which expresses Curie's law

$$\chi = C/T \tag{3-1}$$

with

$$C = \frac{Nm^2}{3k} \tag{3-33}$$

The quantum theory of paramagnetism leads to analogous results, but predicts, in addition, the correct value for **m**.

In the classical theory, **m** is $\sqrt{3}$-times smaller because the spin of the electron is taken as $S = \frac{1}{2}$, instead of $[S(S + 1)]^{1/2} = (\frac{3}{4})^{1/2}$. Consequently, the classical Langevin expression predicts results whose values are one-third those predicted by the quantum mechanical expression, these latter in agreement with the experimental ones.

In the quantum treatment, we have first to establish which are the energy levels of the unpaired electron in the presence of a magnetic field. In order to simplify the problem, we will assume that we deal with a paramagnetic gas containing atoms with one unpaired electron, in a $^2S_{1/2}$ state. Since the atoms of such a gas are sufficiently apart to show no appreciable magnetic interaction, in the absence of a magnetic field they may all be assumed to have the average energy E_0. When the magnetic field is applied, some of the atoms will line up parallel ($m_s = -\frac{1}{2}$) and some antiparallel ($m_s = +\frac{1}{2}$) to the magnetic field. Their new energy levels will be the Zeeman levels of energies

$$E_2 = E_0 + \frac{1}{2} g\mu_B H \tag{3-34}$$

$$E_1 = E_0 - \frac{1}{2} g\mu_B H \tag{3-35}$$

displayed in Fig. 3.1. Therefore, the potential energy difference between levels is

$$E_2 - E_1 = \Delta E = g\mu_B H \tag{3-36}$$

The population ratio of both levels in thermal equilibrium is now determined by H and T

$$N_2/N_1 = \exp(-g\mu_B H/kT) \tag{3-37}$$

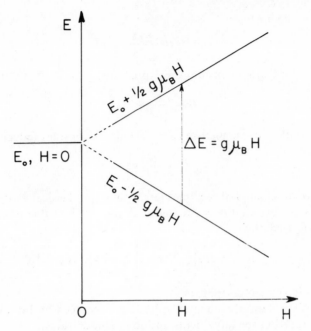

FIG. 3.1 **Zeeman splitting of electronic levels.**

The total magnetization is clearly

$$M = \frac{1}{2} g\mu_B (N_1 - N_2) \tag{3-38}$$

The difference $N_1 - N_2$ is obtained from Eq. (3-37) by replacing $N - N_1$ for N_2

$$N_1 - N_2 = 2N_1 - N = N\left[2\frac{e^x}{1 + e^x} - 1\right] \tag{3-39}$$

where

$$x = g\mu_B H/kT \tag{3-40}$$

For small values of x, Eq. (3-39) simplifies to

$$N_1 - N_2 = N(x/2) = \frac{1}{2} Ng\mu_B H/kT \tag{3-41}$$

and Eq. (3-38) becomes

$$M = \frac{Ng^2\mu_B}{4kT} H \tag{3-42}$$

Hence

$$\chi = M/H = \frac{N}{4kT} g^2 \mu_B^2 \tag{3-43}$$

which gives for Curie's constant the expression

$$C = \frac{N}{4kT} g^2 \mu_B^2 \tag{3-44}$$

For spin-only paramagnetism ($g \approx 2$) Curie's constant reduces to

$$C = \frac{N}{kT} \mu_B^2 \tag{3-45}$$

which is, as anticipated, three times larger than the classical value given by Langevin's expression if m is equated with μ_B. If the value of m given in Eq. (2-19) is used, then

$$m = -2\mu_B \left[\frac{1}{2} \left(\frac{1}{2} + 1 \right) \right]^{1/2} = -\sqrt{3}\mu_B \tag{3-46}$$

and the disagreement disappears.

Langevin's equation, even after m has been replaced by Eq. (3-46), still predicts saturation for large fields and low temperatures. The saturation value, however, becomes $g\mu_B [\frac{1}{2}(\frac{1}{2} + 1)]^{1/2}$, instead of $\frac{1}{2}g\mu_B \approx \mu_B$, as observed. This discrepancy disappears if a more rigorous treatment is made, as follows.

In the case of an atom of total effective spin S,* interaction with the magnetic field will give rise to $2S + 1$ levels corresponding to quantization of the **S** vector along **H** (i.e., $m_S = -S, -(S - 1), \ldots, +S$). The probability that a dipole within the assembly at temperature T has a potential energy

$$E_S = g\mu_B H m_S \tag{3-47}$$

is, according to Boltzmann's statistics,

$$p_S = constant \times \exp(-E_S/kT) \tag{3-48}$$

Here, we assume that the paramagnetic units are sufficiently apart in order not to have any appreciable magnetic interaction among them. Under such conditions, each unit is *distinguishable* and Boltzmann's statistics can be applied. As we will see when dealing with paramagnetism of conduction electrons in a metal, this condition is not always fulfilled and other statistics must be used.

---

*The effective spin S is, in fact, defined by setting the multiplicity of the ground state equal to $2S + 1$ (see Section 9.2).

The constant of Eq. (3-48) is easily found since the summation over all the m_S states must be unity

$$p_S = \frac{\exp(-g\mu_B H m_S/kT)}{\displaystyle\sum_{m_S=-S}^{S} \exp(-g\mu_B N m_S/kT)} \tag{3-49}$$

The total magnetization per unit volume is

$$M = (1/V)\sum_i m_i \tag{3-50}$$

If N is the number of paramagnetic units per unit volume, we can write Eq. (3-50) in terms of the probabilities p_S, in the direction of the magnetic field:

$$M_H = (1/V)N \sum_{m_S=-S}^{S} p_S m_S \tag{3-51}$$

which becomes, after introduction of Eq. (3-49),

$$M_H = \frac{N \displaystyle\sum_{m_S=-S}^{S} -g\mu_B m_S \exp(-g\mu_B H m_S/kT)}{\displaystyle\sum_{m_S=-S}^{S} \exp(-g\mu_B H m_S/kT)} \tag{3-52}$$

Eq. (3-52) may be reduced, after some algebraic handling, to the form

$$M_H = NgS\mu_B \left\{ \frac{2S+1}{2S} \coth\left[\frac{2S+1}{2S} y\right] - \frac{1}{2S} \coth \frac{y}{2S} \right\} = NgS\mu_B B_S(y) \tag{3-53}$$

where

$$y = Sg\mu_B H/kT \tag{3-54}$$

The expression in brackets in Eq. (3-53), $B_S(y)$, is called the Brillouin function. For small values of y, Eq. (3-53) leads to Curie's law. For very large values of H, it predicts saturation with the correct value

$$M_{sat.} = g\mu_B S \tag{3-55}$$

Eq. (3-53) has been thoroughly checked by Henry[3] in three cases: Cr^{3+} in potassium chrome alum ($S = 3/2$), Fe^{3+} in iron ammonium alum ($S = 5/2$), and Gd^{3+} in gadolinium sulfate octahydrate ($S = 7/2$), at temperatures between 1 and 4°K, and at magnetic fields up to 40,000 oersted.

In all cases, the agreement of the experimental results with the theoretical values predicted by the Brillouin function was remarkable.

3.4 Effective number of magnetons

Curie's law is usually written

$$\chi = N_A p_{eff.}^2 \, \mu_B^2 / 3kT \tag{3-56}$$

where N_A is the Avogadro constant and

$$p_{eff.} = g[S(S + 1)]^{1/2} \tag{3-57}$$

is called the *effective number of magnetons*. If one replaces the figures for the constants in Eq. (3-56), the value of p_{eff} is given by

$$p_{eff.} = 2.839 \sqrt{C} \tag{3-58}$$

where $C = \chi T$. It is not unusual in the literature to find results expressed in terms of Weiss magnetons. In this case, the effective number of Weiss magnetons is

$$p_W = 14.05 \sqrt{C} \tag{3-59}$$

but, since the Weiss unit has no theoretical significance, we shall not employ it.

3.5 Paramagnetism of conduction electrons

Contrary to what could be expected at first sight, conductivity electrons in a metal do not follow Curie's law. In this case, the classical treatment, which leads to Langevin's equation, fails completely from two points of view. On one hand, it predicts an effect that varies with $1/T$, while paramagnetism of conduction electrons is temperature independent. On the other hand, it predicts, at room temperature, an effect about 10^2 times larger than the observed one. Although it would be out of the scope of this book to treat this problem in detail, it seems advisable to develop a qualitative argument that provides the right answer, at least within the order of magnitude.

The interpretation of Langevin's expression, derived in Section 3.3, is that the probability of an unpaired electron of aligning parallel to the magnetic field exceeds the probability of aligning antiparallel by a factor of the order of mH/kT. For N atoms, this gives a net magnetization *ca.* (Nm^2H/kT). But this result is based on the implicit assumption of existence of as many empty states as necessary, eventually to reach saturation. This is not the case of conduction electrons in a metal. In applying Fermi-Dirac, instead of Maxwell-Boltzmann statistics, it is found that in the case of metals only a fraction T/T_F of the conduction electrons will find an empty state to line up parallel to the magnetic field. Therefore, Langevin's expression must be multiplied by T/T_F to obtain the actual susceptibility, with the result

$$\chi \approx (Nm^2/kT)\cdot(T/T_F) = Nm^2/kT_F \qquad (3\text{-}60)$$

The Fermi temperature T_F is defined by

$$E_F(0) = kT_F \qquad (3\text{-}61)$$

where $E_F(0)$, called the Fermi energy, is the energy of the level at absolute zero, below which all the states are filled, and above which all the states are empty. Since T_F is of the order of 10^4 to $10^5\,°K$, Eq. (3-60) is not only temperature independent but also predicts the correct order of magnitude of the observed susceptibilities.

References

1. Brindley, G. W., *Phil. Mag.* **11**, 786 (1931).
2. Hartree, D. R., *Repts. Progr. Phys.* **11**, 113 (1946–47).
3. Henry, W. E., *Phys. Rev.* **88**, 559 (1952).
4. Myers, W. R., *Rev. Mod. Phys.* **24**, 15 (1952).
5. Sidwell, T. W., and Hurst, R. P., *J. Chem. Phys.* **37**, 203 (1962).
6. Slater, J. C., *Phys. Rev.* **36**, 57 (1930).
7. Starr, C., and Kaufman, A. R., *Phys. Rev.* **59**, 476 (1941).
8. Stoner, E. C., *Proc. Leeds Phil. Soc. Sci. Sect.* **1**, 484 (1929).
9. Wills, A. P., and Hector, L. G., *Phys. Rev.* **23**, 209 (1924).

General References

Bleaney, B. I., and Bleaney, B., "Electricity and Magnetism," Oxford, Clarendon Press, 1957.
Kittel, C., "Introduction to Solid State Physics," 2nd. ed., New York, John Wiley & Sons, 1960.
Van Vleck, J. H., "The Theory of Electric and Magnetic Susceptibilities," New York, Oxford University Press, 1932.

4 *Thermodynamics of magnetism*

4.1 Magnetic work

The combined expression of the First and Second Laws of Thermodynamics may be written as

$$dU = TdS - PdV - dW \tag{4-1}$$

where U is the internal energy, S the entropy, T the absolute temperature, P the pressure, V the volume, and W the work (other than volume work). The extensive variables (U, S, V, ...) refer to one mole of substance, unless otherwise stated. The work W is defined as positive when done by the system against the external forces. In the case of polarizable media, the work W is expressed by two coupled variables: momentum and field. In dealing with magnetic polarizability, the coupled variables are the total magnetic moment $\mathbf{M}$ and the intensity of the magnetic field. This latter is the intensive variable; the former, the extensive one. Instead of products like PV or TS the scalar product $\mathbf{H} \cdot \mathbf{M}$ should be used. In order to simplify the treatment, this study will be restricted to the isotropic case, in which

$$\mathbf{H} \cdot \mathbf{M} = HM \qquad (4\text{-}2)$$

In magnetizing the substance, work is done by the field, which may be written

$$dW = -HdM \qquad (4\text{-}3)$$

Hence, Eq. (4-1) becomes

$$dU = TdS - PdV + HdM \qquad (4\text{-}4)$$

if there are only volume and magnetic works. Since, at constant temperature and pressure

$$dU + PdV - TdS = d(U + PV - TS) = dG \qquad (4\text{-}5)$$

where G is Gibb's free energy, we have, at constant temperature and pressure:

$$dG = HdM \qquad (4\text{-}6)$$

Notice that we have tacitly assumed one-to-one correspondence between M and H at a given temperature; otherwise Eq. (4-6) would not define a thermodynamic potential. Such a correspondence does, in fact, exist in diamagnetic and paramagnetic substances, for which the equations of state

$$\text{DIAMAGNETISM:} \quad M = H \times constant \qquad (4\text{-}7)$$

$$\text{PARAMAGNETISM:} \quad MT = CH \qquad (4\text{-}8)$$

hold. Eqs. (4-7) and (4-8) have been derived in Sections 3.2 and 3.3, respectively. Ferromagnetism must be excluded from this treatment since, due to hysteresis, such a correspondence does not exist. Antiferromagnetic and ferromagnetic substances, however, follow an equation of state above the transition point. Therefore, they can be included in the general treatment provided that the appropriate equation of state be used, which in the general case may be expressed as

$$f(M, H, T) = 0 \qquad (4\text{-}9)$$

The same holds for paramagnetism of conduction electrons.

Returning to paramagnetic substances that follow Curie's law, the present treatment may be applied to isotropic samples* and single crystals

---

*Isotropic samples may be (i) gases (ii) liquids (iii) isotropic crystals (iv) glasses (v) polycrystalline solids or (vi) powdered crystals, randomly oriented. In the latter two cases, however, if the anisotropy is large, the equation of state may not be as simple as Curie's law.

oriented in such a way that **H** is parallel to one of the principal magnetic axes (Chapter 7).

4.2 Thermodynamic relations

Several relations may be derived by following usual thermodynamic procedures.

At constant temperature and pressure, Eq. (4-6) leads to

$$\left(\frac{\partial G}{\partial M}\right)_{P,T} = H \tag{4-10}$$

At constant volume and temperature Helmholtz's thermodynamic potential F may be introduced, and then

$$dF = d(U - TS) = HdM \tag{4-11}$$

The relations

$$\left(\frac{\partial F}{\partial M}\right)_{T,V} = H \tag{4-12}$$

$$\left(\frac{\partial U}{\partial M}\right)_{S,V} = H \tag{4-13}$$

$$\left(\frac{\partial S}{\partial M}\right)_{U,V} = -\frac{H}{T} \tag{4-14}$$

follow immediately.

At constant entropy and pressure an expression relating the enthalpy H and the magnetic work is obtained:

$$d\mathrm{H} = d(U + PV) = HdM \tag{4-15}$$

In order to be consistent with current thermodynamic notation roman capital H is used for enthalpy and italic capital H for magnetic field. Care should be taken not to exchange their meaning.

From Eq. (4-15) it follows

$$\left(\frac{\partial \mathrm{H}}{\partial M}\right)_{S,P} = H \tag{4-16}$$

In condensed phases, it is justified to neglect the volume work. Hence

$$dU = TdS + HdM \tag{4-17}$$

Now Eq. (4-17) can be treated in order to derive the magnetic Maxwell relations, as follows.

If we take any independent variables x and y and use the simplified symbolism

$$\frac{\partial}{\partial x} = \left(\frac{\partial}{\partial x}\right)_y, \quad \frac{\partial}{\partial y} = \left(\frac{\partial}{\partial y}\right)_x \tag{4-18}$$

we have, from Eq. (4-17):

$$\frac{\partial U}{\partial x} = T\frac{\partial S}{\partial x} + H\frac{\partial M}{\partial x} \tag{4-19}$$

$$\frac{\partial U}{\partial y} = T\frac{\partial S}{\partial y} + H\frac{\partial M}{\partial y} \tag{4-20}$$

Further partial differentiation of Eq. (4-19), with respect to y at constant x, and of Eq. (4-20), with respect to x at constant y, leads to

$$\frac{\partial}{\partial y}\left(\frac{\partial U}{\partial x}\right) = \left(\frac{\partial T}{\partial y}\right)\left(\frac{\partial S}{\partial x}\right) + T\frac{\partial}{\partial y}\left(\frac{\partial S}{\partial x}\right) + \left(\frac{\partial H}{\partial y}\right)\left(\frac{\partial M}{\partial x}\right) + H\frac{\partial}{\partial y}\left(\frac{\partial M}{\partial x}\right) \tag{4-21}$$

and to

$$\frac{\partial}{\partial x}\left(\frac{\partial U}{\partial y}\right) = \left(\frac{\partial T}{\partial x}\right)\left(\frac{\partial S}{\partial y}\right) + T\frac{\partial}{\partial x}\left(\frac{\partial S}{\partial y}\right) + \left(\frac{\partial H}{\partial x}\right)\left(\frac{\partial M}{\partial y}\right) + H\frac{\partial}{\partial x}\left(\frac{\partial M}{\partial y}\right) \tag{4-22}$$

Since dU, dS, and dM are all perfect differentials, the operators $\frac{\partial}{\partial x}$ and $\frac{\partial}{\partial y}$ commute, and then

$$\frac{\partial}{\partial y}\left(\frac{\partial U}{\partial x}\right) = \frac{\partial}{\partial x}\left(\frac{\partial U}{\partial y}\right) \tag{4-23}$$

$$\frac{\partial}{\partial y}\left(\frac{\partial S}{\partial x}\right) = \frac{\partial}{\partial x}\left(\frac{\partial S}{\partial y}\right) \tag{4-24}$$

$$\frac{\partial}{\partial y}\left(\frac{\partial M}{\partial x}\right) = \frac{\partial}{\partial x}\left(\frac{\partial M}{\partial y}\right) \tag{4-25}$$

After taking into account Eqs. (4-23) to 4-25), Eqs. (4-21) and (4-22) reduce to

$$\left(\frac{\partial T}{\partial y}\right)\left(\frac{\partial S}{\partial x}\right) + \left(\frac{\partial H}{\partial y}\right)\left(\frac{\partial M}{\partial y}\right) = \left(\frac{\partial T}{\partial x}\right)\left(\frac{\partial S}{\partial y}\right) + \left(\frac{\partial H}{\partial x}\right)\left(\frac{\partial M}{\partial y}\right) \tag{4-26}$$

We now substitute for the independent variables x and y any uncoupled pair* taken from S, T, M, and H. There are four such combinations:

$$x = S, y = M : \left(\frac{\partial T}{\partial M}\right)_S = \left(\frac{\partial H}{\partial S}\right)_M \qquad (4\text{-}27)$$

$$x = S, y = H : \left(\frac{\partial T}{\partial H}\right)_S = \left(\frac{\partial M}{\partial S}\right)_H \qquad (4\text{-}28)$$

$$x = T, y = H : \left(\frac{\partial M}{\partial T}\right)_H = \left(\frac{\partial S}{\partial H}\right)_T \qquad (4\text{-}29)$$

$$x = T, y = M : \left(\frac{\partial S}{\partial M}\right)_T = \left(\frac{\partial H}{\partial T}\right)_M \qquad (4\text{-}30)$$

Eqs. (4-27) to (4-30) are the magnetic analogs for Maxwell's well-known thermodynamic relations involving S, T, V, and P.

Eq. (4-29) is especially useful in magnetic cooling since it shows that, if a sample has a magnetic susceptibility which *decreases with increasing temperature*, the entropy of the system will be lowered on isothermal magnetization, heat being transferred from the sample to the surroundings. Due to the positive sign of the specific heats (in regions where neither first- nor second-order transitions take place), adiabatic magnetization *must increase* the temperature of the sample. In order to find the rise in temperature, one may write

$$T dS = C_H dT + T \left(\frac{\partial M}{\partial T}\right)_H dH \qquad (4\text{-}31)$$

where

$$C_H = T \left(\frac{\partial S}{\partial T}\right)_H \qquad (4\text{-}32)$$

is the specific heat measured at constant magnetic field. Under adiabatic conditions, $dS = 0$, one arrives at

$$\Delta T = - \int_0^H (T/C_H) \left(\frac{\partial M}{\partial T}\right)_H dH \qquad (4\text{-}33)$$

Clearly, $\Delta T = 0$ in diamagnetic and temperature independent paramagnetic samples, and negative if the magnetic susceptibility *increases with increasing temperature*.

---

*Uncoupled pairs are readily recognized by the fact that their product does not have units of energy. Coupled pairs, on the contrary, multiply to give an energy term, e.g.: TS, PV, MH, and are not independent variables.

4.3   Thermodynamics of magnetic cooling

Since magnetization involves work done on the specimen by external forces, if it is done isothermally and the specimen then adiabatically demagnetized, the total magnetic work done on the specimen during isothermal magnetization will be realized by the specimen during the adiabatic demagnetization, with consequent lowering of the specimen temperature. This is the basis of magnetic cooling, with which temperatures as low as 10^{-3} °K may be reached demagnetizing substances with unpaired electrons, and as low as 10^{-6} °K using nuclear demagnetization. In order to treat the thermodynamics of the process of magnetic cooling, we proceed by writing the expression at constant volume

$$dU = T\,dS + H\,dM \qquad (4\text{-}34)$$

If the substance is paramagnetic, there is an equation of state that relates H, M, and T. Therefore, it can be established that both the entropy and the magnetization are functions of T and H

$$S = S(T, H) \qquad (4\text{-}35)$$

$$M = M(T, H) \qquad (4\text{-}36)$$

Eq. (4-34) may be written as

$$dU = T\left[\left(\frac{\partial S}{\partial T}\right)_H dT + \left(\frac{\partial S}{\partial H}\right)_T dH\right] + H\left[\left(\frac{\partial M}{\partial T}\right)_H dT + \left(\frac{\partial M}{\partial H}\right)_T dH\right] \qquad (4\text{-}37)$$

Since the internal energy U is a function of the state of the system, dU is an exact differential. Hence

$$dU = \left(\frac{\partial U}{\partial T}\right)_H dT + \left(\frac{\partial U}{\partial H}\right)_T dH \qquad (4\text{-}38)$$

must be equal to Eq. (4-37). Therefore,

$$\left(\frac{\partial U}{\partial T}\right)_H = T\left(\frac{\partial S}{\partial T}\right)_H + H\left(\frac{\partial M}{\partial T}\right)_H \qquad (4\text{-}39)$$

and

$$\left(\frac{\partial U}{\partial H}\right)_T = T\left(\frac{\partial S}{\partial H}\right)_T + H\left(\frac{\partial M}{\partial H}\right)_T \qquad (4\text{-}40)$$

Taking into account Eq. (4-29) we arrive at

$$\left(\frac{\partial U}{\partial H}\right)_T = T\left(\frac{\partial M}{\partial T}\right)_H + H\left(\frac{\partial M}{\partial H}\right)_T \qquad (4\text{-}41)$$

The magnetization M is a function of T and H, and dM (as in the case of the internal energy) is an exact differential. The magnetic term of Eq.

(4-34) may then be expressed as

$$H\,dM = H\left(\frac{\partial M}{\partial T}\right)_H dT + H\left(\frac{\partial M}{\partial H}\right)_T dH \qquad (4\text{-}42)$$

Combining Eqs. (4-34), (4-38), and (4-42), one obtains

$$T\,dS = \left(\frac{\partial U}{\partial H}\right)_T dH + \left(\frac{\partial U}{\partial T}\right)_H dT - H\left(\frac{\partial M}{\partial H}\right)_T dH - H\left(\frac{\partial M}{\partial T}\right)_H dT \qquad (4\text{-}43)$$

which, after arranging differentials; gives:

$$T\,dS = \left[\left(\frac{\partial U}{\partial T}\right)_H - H\left(\frac{\partial M}{\partial T}\right)_H\right] dT + \left[\left(\frac{\partial U}{\partial H}\right)_T - H\left(\frac{\partial M}{\partial H}\right)_T\right] dH \qquad (4\text{-}44)$$

Differentiation of Eq. (4-34) at constant temperature results in

$$T\left(\frac{\partial S}{\partial H}\right)_T = \left(\frac{\partial U}{\partial H}\right)_T - H\left(\frac{\partial M}{\partial H}\right)_T \qquad (4\text{-}45)$$

which, introduced into Eq. (4-44), after taking into account Eq. (4-29), finally leads to

$$T\,dS = \left[\left(\frac{\partial U}{\partial T}\right)_H - H\left(\frac{\partial M}{\partial T}\right)_H\right] dT + T\left(\frac{\partial M}{\partial T}\right)_H dH \qquad (4\text{-}46)$$

Eq. (4-46) is important since it permits the evaluation of the heat Q supplied to the specimen during isothermal magnetization:

$$Q = T\Delta S = T\int_0^H \left(\frac{\partial M}{\partial T}\right)_H dH \qquad (4\text{-}47)$$

from which follows the entropy change

$$S(H, T) - S(0, T) = \int_0^H \left(\frac{\partial M}{\partial T}\right)_H dH \qquad (4\text{-}48)$$

Clearly, the entropy will decrease during isothermal magnetization if the magnetization decreases with increasing temperature, for in such a case the integrand of Eq. (4-48) is negative.

In the case of a substance that follows Curie's law

$$M = (C/T)\,H \qquad (4\text{-}48)$$

one has

$$\left(\frac{\partial M}{\partial T}\right)_H = -\frac{C}{T^2}\,H \qquad (4\text{-}49)$$

Introducing Eq. (4-49) into Eq. (4-48) and taking into account that

$$\left(\frac{\partial M}{\partial H}\right)_T = \frac{C}{T} \tag{4-50}$$

Eq. (4-48) may be written

$$S(H, T) = S(0, T) - \frac{1}{T} \int_0^H H \, dM \tag{4-51}$$

Eq. (4-51) means that the decrease in entropy of a substance, due to isothermal magnetization, is equal to the external magnetic work divided by the temperature of magnetization. Thus, heat is liberated by the sample during isothermal magnetization and the entropy subsequently lowered. The heat liberated by the sample is transferred to the thermostat. If the thermal contact is now broken and the magnetic field is switched off, the entropy remains constant and the temperature of the sample must decrease from T_i to T_f according to

$$S(0, T_f) = S(H, T_i) \tag{4-52}$$

where T_i is obviously the temperature at which magnetization was performed.

Recalling the statistical meaning of the entropy, the situation may be described in the following terms. During isothermal magnetization, disorder is partially removed by the effect of the magnetic field in aligning the magnetic dipoles. The new state of partial order cannot change during any adiabatic process, since such a process must be isoentropic. If the reason for aligning disappears, the temperature of the sample must fall to the value at which this state of partial order is compatible with thermal agitation.

In order to estimate the drop in temperature, we proceed as follows. Since dS is an exact differential

$$T \, dS = T \left(\frac{\partial S}{\partial T}\right)_H dT + T \left(\frac{\partial S}{\partial H}\right)_T dH \tag{4-53}$$

where H and T have been taken as the independent variables. In adiabatic (isoentropic) demagnetization Eq. (4-53) reduces to

$$T \left(\frac{\partial S}{\partial T}\right)_H dT = - T \left(\frac{\partial S}{\partial H}\right)_T dH \tag{4-54}$$

Introducing Maxwell's relation (4-29) and the specific heat at constant magnetic field defined in Eq. (4-32), one obtains

$$C_H dT = - T \left(\frac{\partial M}{\partial T}\right)_H dH \tag{4-55}$$

Now, from the definition of C_H it follows

$$\left(\frac{\partial T}{\partial S}\right)_H = \frac{T}{C_H} \tag{4-32}$$

while

$$\left(\frac{\partial M}{\partial T}\right)_H \left(\frac{\partial T}{\partial S}\right)_H = \left(\frac{\partial M}{\partial S}\right)_H \tag{4-56}$$

Multiplying both members of Eq. (4-55) by $\left(\frac{\partial T}{\partial S}\right)_H$ and substituting Eqs. (4-32) and (4-56), one arrives at the change in temperature following adiabatic magnetization from H_1 to H_2

$$T_2 - T_1 = - \int_{H_1}^{H_2} \left(\frac{\partial M}{\partial S}\right)_H dH \tag{4-57}$$

The decrease in entropy during magnetization can be found by considering M and T as the independent variables, therefore writing

$$T dS = T\left(\frac{\partial S}{\partial T}\right)_M dT + T\left(\frac{\partial S}{\partial M}\right)_T dM \tag{4-58}$$

As we have defined C_H in Eq. (4-32), the specific heat at constant magnetization can now be defined

$$C_M = T\left(\frac{\partial S}{\partial T}\right)_M \tag{4-59}$$

Then, introducing Eq. (4-59) and Maxwell's relation (4-30) in Eq. (4-58) one arrives at

$$T dS = C_M dT - T\left(\frac{\partial H}{\partial T}\right)_M dM \tag{4-60}$$

which, for isothermal magnetization, reduces to

$$T dS = -T\left(\frac{\partial H}{\partial T}\right)_M dM \tag{4-61}$$

Integration of Eq. (4-61) gives the entropy change

$$\Delta S = - \int_0^H \left(\frac{\partial H}{\partial T}\right)_M dM \tag{4-62}$$

Clearly, the entropy change of Eq. (4-62) may be expressed

$$\Delta S = S(H = H_f, T) - S(H = 0, T) \tag{4-63}$$

for there is a function of state $f(H, M, T) = 0$ which establishes, at constant temperature, a one-to-one correspondence between magnetization M and field H.

In the case of ideal paramagnetic behavior, Eq. (4-9) takes the explicit form of Curie's law, from which

$$\left(\frac{\partial H}{\partial T}\right)_M = \frac{M}{C} \tag{4-64}$$

follows. The entropy change may then be expressed as

$$\Delta S = \frac{1}{C} \int_0^H M \, dM = \frac{1}{2C} M^2 \tag{4-65}$$

or

$$\Delta S = \frac{1}{2} (C/T^2) H^2 \tag{4-66}$$

or

$$\Delta S = \frac{1}{2} (M/T) H \tag{4-67}$$

Eqs (4-65) to (4-67) being clearly equivalent.

4.4   The paramagnetic thermometer

Conventional thermometers—based in measuring the pressure of a gas at constant volume, the resistivity change of an electric conductor, the vapor pressure of a suitable liquid in equilibrium with its vapor, and the electromotive force of a thermocouple—loose sensitivity at low temperatures, rapidly becoming unuseful below 1°K. It is then necessary to resort to measurement of the magnetization of a paramagnetic salt in a small, well-known magnetic field. This method provides a convenient secondary thermometer since the magnetization is, in general, very large at very low temperatures and increases as the temperature falls. Measurement of the susceptibility is performed by the mutual inductance method, described in Section 6.7, over samples shaped to a sphere. Otherwise, if the samples have other shapes (as, for example, ellipsoids of revolution) the measured susceptibility is corrected to that of a sphere by suitable formulas. Unfortunately, these corrections are not free of objections, especially at very low temperatures (of some thousandths of a degree) where cooperative effects may set in. In addition, at these low temperatures and at the frequencies used in the mutual inductance methods, there may appear a significant lag between the magnetization and the applied oscillating field due to comparatively long paramagnetic relaxation times. If these objections are not considered, the paramagnetic thermometer provides a measure of the temperature which we call T^*, directly derived from Curie's law. A discussion as to the extent that this scale agrees with the absolute thermodynamic temperature follows.

Translation of the paramagnetic temperature T^* into the absolute temperature T may be performed in several ways. A commonly used method is first to obtain the curve S *vs.* T^* for $H = 0$ by a series of adiabatic demagnetizations from different fields to $H = 0$, starting each time from 1°K. The entropy value at this temperature is evaluated from the theoretical expressions of the magnetic susceptibility derived in Section 3.3. Next, the amount of heat, Q, needed to warm the sample from each value of T^* to the initial temperature of 1°K is plotted as a function of T^*. The heating is usually done by irradiation of the sample with gamma rays from a radioactive source. The absolute amount of heat supplied must be obtained by calibration with a well-known substance at 1°K, assuming that both scales of temperature coincide at 1°K. The correspondence between T and T^* is then given by

$$T = \frac{\Delta Q}{\Delta S} = \frac{\left(\dfrac{\partial Q}{\partial T^*}\right)_{H=0}}{\left(\dfrac{\partial S}{\partial T^*}\right)_{H=0}} \qquad (4\text{-}6)$$

The scale of paramagnetic temperature clearly depends on the substance whose magnetization is measured. In the case of iron ammonium alum, the two scales do not differ greatly down to 0.3°K. Below this temperature, however, the departure of T^* from T increases until $T^* \approx 0.07°$ ($T = 0.04°$K), where hysteresis phenomena begin to appear. Chromium potassium alum, for example, provides $T^* = 0.06°$ for $T = 0.035°$K and $T^* = 0.033°$ for $T = 0.0039°$K!

4.5   Limitations of electron paramagnetic cooling

In an ideal paramagnetic gas (pure spins with no interaction), the initial (T_i) and final (T_f) temperatures during demagnetization from H_i to H_f would satisfy the relation

$$T_f H_i = T_i H_f \qquad (4\text{-}69)$$

In practice, the value of H_f cannot be zero, for the effective magnetic field $\mathbf{H}_{eff.}$ acting on an unpaired electron is

$$\mathbf{H}_{eff.} = \mathbf{H} + \mathbf{h} \tag{4-70}$$

where $\mathbf{H}$ is the external field and $\mathbf{h}$ is the internal magnetic field due to magnetic interactions of the type studied in Section 1.6. Eq. (4-69) then takes the form

$$T_f(H_i^2 + h^2 + 2H_i h \cos\theta)^{1/2} = T_i(H_f^2 + h^2 + 2H_f h \cos\theta)^{1/2} \tag{4-71}$$

where θ is the angle between $\mathbf{H}$ and $\mathbf{h}$.

If demagnetization is performed by switching off the external field, $H_f = 0$. Since, in addition, the initial field H_i is much larger than h, Eq. (4-71) may be simplified to

$$T_f H_i = T_i h \tag{4-72}$$

This relation limits the lowest attainable temperature to a value T_L such that

$$hm \approx kT_L \tag{4-73}$$

where m is the average magnetization of the atom, ion, or molecule (see Sections 2.1 and 2.5). Since h is of the order m/r^3 (see Section 1.6) the limiting temperature is of the order

$$T_L \approx mh/k \approx (1/k)(m^2/r^3) \tag{4-74}$$

In conveniently diluted paramagnetic samples* it is possible to reach temperatures of the order of $10^{-3}\,^\circ K$. Less diluted samples would have larger values of h, therefore being less efficient. In order to obtain temperatures significantly below this limit, it is necessary to resort to nuclear demagnetization, for in this case m is about three orders of magnitude smaller than the electron value, and so is h. Since the method of nuclear demagnetization is out of the scope of this book; it will suffice to say that by nuclear demagnetization temperatures as low as $10^{-6}\,^\circ K$ have been reached.

General References

Casimir, H. B. G., "Magnetism and Very Low Temperatures," New York, Dover Publications, 1961.

*A paramagnetic sample is said to be dilute when the paramagnetic units are far enough from one another not to involve appreciable magnetic interaction. In some cases, dilution may be accomplished by growing mixed crystals of the paramagnetic compound and an isomorphous, diamagnetic salt, e.g., chromium aluminum alum with Cr:Al ratios ca. 1:15. In other cases, salts of very large diamagnetic anions are used, e.g., $Gd(PMo_{12}O_{40}) \cdot 30H_2O$, in which, in addition, the distribution of the paramagnetic ion is uniform.

Garrett, C. G. B., "Magnetic Cooling," Cambridge, Harvard University Press, 1954.

Gorter, C. J., "Paramagnetic Relaxation," Amsterdam, Elsevier Publishing Company, 1947.

Gorter, C. J., "Progress in Low Temperature Physics," Vol. 1, Amsterdam, North-Holland Publishing Company, 1957.

Guggenheim, E. A., "Thermodynamics," New York, Interscience Publishers, 1949.

Jackson, L. C., "Low Temperature Physics," Fourth Ed., London, Methuen and Co., 1955.

Rossini, F. D., "Chemical Thermodynamics," New York, John Wiley & Sons, Inc., 1950.

5   Symmetry of molecules
       and crystals

5.1   Elements and operations of symmetry

Symmetry is a property inherent in geometrical shapes; it may be (and indeed, is) extrapolated to abstract concepts. Perhaps the most familiar case of symmetry is the correlation between an object and its image produced by a plane mirror. This correlation may be studied in a simple way by a convenient choice of the reference frame. Consider, as illustrated in Fig. 5.1, an orthogonal system of coordinates (x, y, z) oriented in such a way that the mirror plane, σ, lies in the plane (y, z) intersecting the x-axis at $x = 0$. The correlation between object and image may be treated as follows: to each object point P of coordinates x, y, z there corresponds an image point P' of coordinates $-x, y, z$. This example shows that symmetry involves two concepts: an *operation* that transforms a point into an equivalent point, in this case by changing the sign of x, and a geometrical entity about which the operation is performed. This entity is *unchanged by the operation of symmetry*. In the example discussed, the operation is a

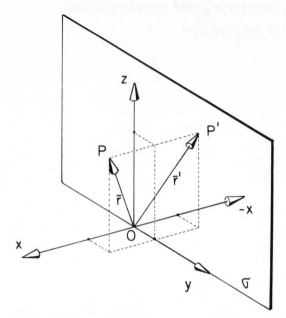

FIG. 5-1   Symmetry operation of reflection.

reflection, the element, a *symmetry plane.* In regard to the symmetry operation, there is no difference between object and image other than that established by the symmetry correlation. Their points all belong to the same configuration, it being irrelevant to ask which part of the configuration is the object and which the image, for each one transforms into the other by the same operation, namely: *changing the sign of x.* Symmetry planes are often found in crystals and molecules, as illustrated in Fig. 5.2.

Another type of symmetry element is the *symmetry axis,* about which *n-fold rotations* are performed. An *n*-fold rotation is a rotation by $2\pi/n$ and integral multiples of such an angle. Thus, 2-fold, 3-fold, 4-fold,... rotations are rotations by π, $2\pi/3$, $\frac{1}{2}\pi$,.... Corresponding to each of them is an element of symmetry called *n-fold axis.* Fig. 5.3 shows a 2-fold axis in a crystal and a 3-fold axis in a molecule.

There is an element of symmetry called *inversion center* or *center of symmetry,* about which the whole configuration is inverted. The operation is called *inversion.* If the origin of the reference frame is brought to coincide with such a center, the operation changes the sign of all three coordinates. Fig. 5.4 shows a body (cube) that has a symmetry center *i,* and a regular tetrahedron (in solid lines) that has no center of symmetry. The absence of a center of symmetry in the tetrahedron is evident if one per-

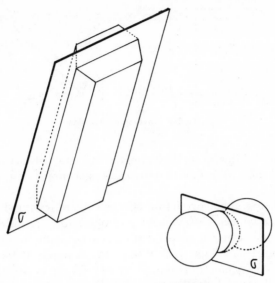

FIG. 5-2 Planes of symmetry.

forms the inversion operation about the center of figure f; in such a case, the tetrahedron shown in dotted lines is obtained. Since the inverted tetrahedron does not overlap the unoperated figure, the inversion operation is not a symmetry operation of the tetrahedron.

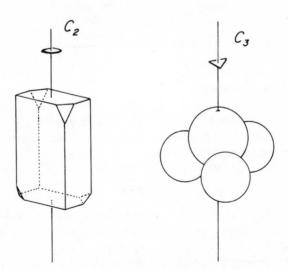

FIG. 5-3 Symmetry axes of proper rotations.

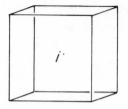
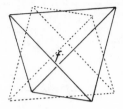

FIG. 5-4   Inversion operation.

There is a trivial operation of symmetry that may be performed in all bodies: the *identity operation* that replaces each point of the configuration *by itself.*

Table 5.1 displays the correlation between operations and elements of symmetry. The concepts of element and operation are different in that although only one element corresponds to each operation, the reciprocal statement is not true. To a six-fold axis, for example, there correspond five operations of symmetry (rotation by $\pi/3$, $2\pi/3$, π, $4\pi/3$, and $5\pi/3$) in addition to the trivial identity rotation by 2π.

TABLE 5.1   Correlation between operations and elements of symmetry.

Element	Operation
plane	reflection
axis	rotation
center	inversion

5.2   Symmetry operation

Definition: *An operation of symmetry is a mathematical artifice that changes each point of a configuration into an equivalent point of the same configuration, and eventually into itself.*

An operation of symmetry may be expressed by the symbolic equation

$$\mathbf{r}' = \mathbf{R} \cdot \mathbf{r} \qquad (5\text{-}1)$$

where $\mathbf{r}$ and $\mathbf{r}'$ are vectors representing two equivalent points in an arbitrary frame. The properties of $\mathbf{R}$, an operator, will be discussed below. The difference between a symmetry operator and a symmetry operation is formal. The symmetry operator is the symbol $\mathbf{R}$, while the symmetry operation is represented by Eq. (5-1). There is one-to-one correspondence between symmetry operators and symmetry operations. Therefore, referral to either one will not cause misunderstanding.

In order to derive the general properties of the **R** operator, it is not necessary to assign it any explicit formulation. The definition of the **R** operator as an artifice that changes one vector into another, vague as it may seem, is sufficient for what follows. The reason for introducing such a general definition arises from the fact that the explicit formulation of the symmetry operator depends on the reference frame. In Fig. 5.5, for example, the reflection operation is referred to two frames. In case (a)

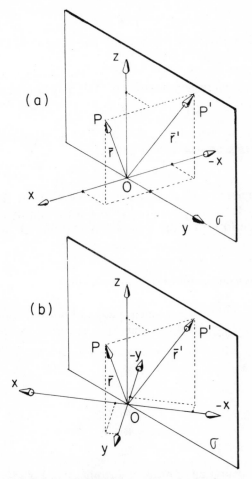

FIG. 5-5 Symmetry operation of reflection referred to two different systems of coordinates.

the operation is performed by changing the sign of x. In case (b), the same operation must be performed by changing x into $-y$ and y into $-x$. Less fortunate choices of the reference frame may lead to more complex formulations of the **R** operator.

5.3   Association of symmetry operators

Let **r**, **r'**, and **r''** be equivalent points, and **R** and **R'** symmetry operators defined by Eq. (5-1) and

$$\mathbf{r}'' = \mathbf{R}' \cdot \mathbf{r}' \tag{5-2}$$

respectively. Elimination of **r'** between Eq. (5-1) and (5-2) leads to

$$\mathbf{r}'' = \mathbf{R}' \cdot \mathbf{R} \cdot \mathbf{r} \tag{5-3}$$

Since **r** and **r''** are equivalent points, Eq. (5-3) may also be written

$$\mathbf{r}'' = \mathbf{R}'' \cdot \mathbf{r} \tag{5-4}$$

where **R''** is another symmetry operator. The product, or multiplication, of symmetry operators can then be defined as

$$\mathbf{R}'' = \mathbf{R}' \cdot \mathbf{R} \tag{5-5}$$

meaning that operation by **R**, followed by operation by **R'**, is equivalent to operation by **R''**. Since the product is not necessarily commutative, it will always be performed in the same way (i.e., from right to left).

5.4   Associative character of the multiplication

Letting $\mathbf{r}_i$, $\mathbf{r}_j$, $\mathbf{r}_k$, and $\mathbf{r}_l$ be equivalent points of a given configuration, and writing the symmetry operations that exchange them, we have

$$\mathbf{r}_j = \mathbf{R}_{ji} \cdot \mathbf{r}_i \tag{5-6}$$
$$\mathbf{r}_k = \mathbf{R}_{kj} \cdot \mathbf{r}_j \tag{5-7}$$
$$\mathbf{r}_l = \mathbf{R}_{lk} \cdot \mathbf{r}_k \tag{5-8}$$
$$\mathbf{r}_k = \mathbf{R}_{ki} \cdot \mathbf{r}_i \tag{5-9}$$
$$\mathbf{r}_l = \mathbf{R}_{lj} \cdot \mathbf{r}_j \tag{5-10}$$
$$\mathbf{r}_l = \mathbf{R}_{li} \cdot \mathbf{r}_i \tag{5-11}$$

By applying the definition of product introduced in Eq. (5-5) we may write

$$\mathbf{R}_{li} = \mathbf{R}_{lj} \cdot \mathbf{R}_{ji} = \mathbf{R}_{lk} \cdot \mathbf{R}_{ki} = (\mathbf{R}_{lk} \cdot \mathbf{R}_{kj}) \cdot \mathbf{R}_{ji} = \mathbf{R}_{lk} \cdot (\mathbf{R}_{kj} \cdot \mathbf{R}_{ji}) \tag{5-12}$$

which shows the associative character of multiplication. Notice that the definition of equality used is: two operators $\mathbf{R}'$ and $\mathbf{R}''$ are equal only if

$$\mathbf{R}' \cdot \mathbf{R} = \mathbf{R}'' \cdot \mathbf{R} \qquad (5\text{-}13)$$

If both members are multiplied by $\mathbf{R}$, we have

$$\mathbf{R} \cdot \mathbf{R}' \cdot \mathbf{R} = \mathbf{R} \cdot \mathbf{R}'' \cdot \mathbf{R} \qquad (5\text{-}14)$$

and then

$$\mathbf{R} \cdot \mathbf{R}' = \mathbf{R} \cdot \mathbf{R}'' \qquad (5\text{-}15)$$

which is an alternative condition of equality.

5.5   Symmetry operators as a group

The product of two symmetry operators is a symmetry operator, since it transforms equivalent points of a given configuration. The set or collection of symmetry operators is therefore said to be a closed set with respect to the law of association (defined in eq. (5-5)).

In each set of symmetry operators there is the trivial identity operator $\mathbf{E}$ that transforms each point of the configuration into itself. It commutes with all symmetry operators leaving them unchanged, i.e.,

$$\mathbf{E} \cdot \mathbf{R} = \mathbf{R} \cdot \mathbf{E} = \mathbf{R} \qquad (5\text{-}16)$$

Given a symmetry operator $\mathbf{R}_i$, an operator $\mathbf{R}_j$ always exists

$$\mathbf{E} = \mathbf{R}_j \cdot \mathbf{R}_i \qquad (5\text{-}17)$$

The operator $\mathbf{R}_j$ is called the *reciprocal* or *inverse* of the operator $\mathbf{R}_i$, symbolized $\mathbf{R}_i^{-1}$ and conversely. In general, it is written

$$\mathbf{E} = \mathbf{R}^{-1} \cdot \mathbf{R} \qquad (5\text{-}18)$$

The product $\mathbf{R}^{-1} \cdot \mathbf{R}$ is commutative. To prove it, multiply Eq. (5-18) by $\mathbf{R}$ and obtain

$$\mathbf{R} \cdot \mathbf{E} = \mathbf{R} \cdot (\mathbf{R}^{-1} \cdot \mathbf{R}) = (\mathbf{R} \cdot \mathbf{R}^{-1}) \cdot \mathbf{R} = \mathbf{E} \cdot \mathbf{R} \qquad (5\text{-}19)$$

since multiplication is associative. In view of the equality condition, it follows that

$$\mathbf{E} = \mathbf{R}^{-1} \cdot \mathbf{R} = \mathbf{R} \cdot \mathbf{R}^{-1} \qquad (5\text{-}20)$$

Each symmetry operator has only one reciprocal operator. To prove it, let $\mathbf{R}_1$ and $\mathbf{R}_2$ be, by assumption, reciprocal operators of $\mathbf{R}$. Then

$$\mathbf{R}_1 \cdot \mathbf{R} = \mathbf{E} = \mathbf{R}_2 \cdot \mathbf{R} \qquad (5\text{-}21)$$

and hence

$$R_1 = R_2 \tag{5-22}$$

Q.E.D.

So far it has been shown that:

(1) The set of operators of symmetry is closed with respect to a law of association called multiplication.

(2) Multiplication of symmetry operators is associative.

(3) There exists an identity operator so that $E \cdot R = R \cdot E = R$.

(4) To each symmetry operator R there corresponds a reciprocal operator R^{-1} so that $R^{-1} \cdot R = R \cdot R^{-1} = E$.

These four conditions are the group postulates; the complete set of symmetry operators is therefore a group. The importance of being a group lies in the fact that groups have been long and thoroughly investigated by mathematicians before they were needed for the solution of physical problems. All the knowledge previously gained in this field is therefore applicable to the symmetry operations in the benefit of physics and chemistry today.

5.6 Multiplication table

Since the set of operations of symmetry of a given configuration is a group, and hence the product of any two operators must be an operator belonging to the same group, it is useful to make a table showing all the operators together with their products. Such a table is known as the *multiplication table of the group* and permits, given any two operators R_i and R_j, to find the operators $R_i \cdot R_j$ and $R_j \cdot R_i$. A useful property of the multiplication table will be derived after application to a particular case.

In order to apply the treatment developed in preceding sections, we will study the symmetry of the equilateral triangle shown in Fig. 5.6. Although it is supposed throughout that all three triangle vertices are equivalent, they are labeled A, B, and C in order to follow the symmetry operations generated by the elements shown in the upper left triangle. There are four elements of symmetry: three planes, σ_1, σ_2, and σ_3, and one 3-fold axis perpendicular to the triangle, symbolized by the smaller, black triangle. Each symmetry plane generates one operator m_i which is its own reciprocal, since

$$m_i^{-1} \cdot m_i = E = m_i \cdot m_i \tag{5-23}$$

It is clear that performing an even number of a given reflection leaves the configuration unchanged.

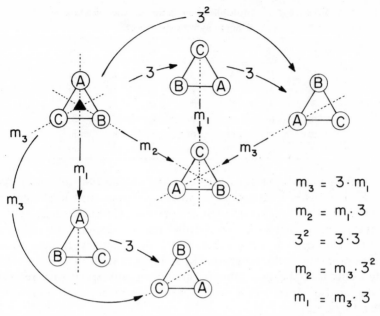

$$m_3 = 3 \cdot m_1$$

$$m_2 = m_1 \cdot 3$$

$$3^2 = 3 \cdot 3$$

$$m_2 = m_3 \cdot 3^2$$

$$m_1 = m_3 \cdot 3$$

FIG. 5-6   Some operations of symmetry of an equilateral triangle.

To the trigonal axis there correspond a rotation by $2\pi/3$, symbolized **3**, a rotation by $4\pi/3$, symbolized 3^2 and equivalent to two **3**-operations:

$$3^2 = 3 \cdot 3 \qquad (5\text{-}24)$$

and finally the trivial identity operation

$$E = 3^3 = 3^2 \cdot 3 = 3 \cdot 3^2 = 3 \cdot 3 \cdot 3 \qquad (5\text{-}25)$$

There are no other symmetry operations. The symmetry group of a two-dimensional, equilateral triangle then has the operators E, m_1, m_2, m_3, **3**, and 3^2. Notice that we have now replaced the symbols E, m_1,... for the general symbol R used in preceding sections. Table 5.2 shows the multiplication table of this group. The upper row and left column display the operators R_i and R_j respectively. The intersection of each row with each column displays the operator $R_j \cdot R_i$ with the multiplication performed, as indicated before, from right to left. In other words, the operator which is to be applied first is written across the top of the table. Some of the products are shown in Fig. (5.6), where one verifies that rotations and reflections do not commute, nor do different reflections. Rotations do commute; the product of two rotations is either a rotation or the identity operator. That the product of two 3-fold rotations is always a 3-fold rota-

TABLE 5.2   Multiplication table of the equilateral triangle symmetry group.

	E	3	3^2	m_1	m_2	m_3
E	E	3	3^2	m_1	m_2	m_3
3	3	3^2	E	m_3	m_1	m_2
3^2	3^2	E	3	m_2	m_3	m_1
m_1	m_1	m_2	m_3	E	3	3^2
m_2	m_2	m_3	m_1	3^2	E	3
m_3	m_3	m_1	m_2	3	3^2	E

tion or the identity operator is expressed by stating that the set of operators $(E, 3, 3^2)$ is in itself a group. Since it belongs to a larger group of symmetry operators, it is said to be a *subgroup*. The reader may verify, as a useful exercise, that such a set satisfies the four group postulates. The identity operator is a trivial subgroup of all symmetry groups. An important property of the subgroup $(E, 3, 3^2)$ is that all the products commute. The multiplication table is therefore symmetrical with respect to the diagonal of operators R_i^2. Groups of operators that commute are called *Abelian*.

5.7   Fundamental property of the multiplication table

All the operators of the group must appear once, and only once, in each row and column of the multiplication table. This property results immediately from considerations that follow. Should an operator appear twice in a row, it would imply that $R_k = R_j \cdot R_i = R \cdot R_i$, and then $R_j = R$, meaning that an operator has been included twice in the table. A similar argument prohibits the appearance of an operator more than once in a column. Since there are as many columns (and rows) as operators, and no operator can appear more than once in each column (or row), each operator *must* appear once in each column (or row). Otherwise the set would not be closed and hence not a group. Verification of this property of the multiplication table is useful, for it may warn that:
(1)   Some operations have been incorrectly performed.
(2)   An operation is missing.
(3)   An operation has been counted twice.

5.8   Formulation of the symmetry operator

Eq. (5-1) essentially represents an operation which is equivalent to a change of coordinates. If (x, y, z) and (x', y', z') are the coordinates of points **r** and **r**' respectively, we may write the linear relations

$$x' = a_{11}x + a_{12}y + a_{13}z$$
$$y' = a_{21}x + a_{22}y + a_{23}z$$
$$z' = a_{31}x + a_{32}y + a_{33}z \tag{5-26}$$

where the coefficients a_{ij} are direction cosines, thus satisfying the conditions

$$\sum_i a_{ij} \cdot a_{ik} = \delta_{jk}$$

$$\sum_i a_{ji} \cdot a_{ki} = \delta_{jk} \tag{5-27}$$

where δ_{jk} is Kronecker's delta.

The set of Eq. (5-26) may be symbolically written as

$$\begin{bmatrix} x' \\ y' \\ z' \end{bmatrix} = \begin{bmatrix} a_{11} & a_{12} & a_{13} \\ a_{21} & a_{22} & a_{23} \\ a_{31} & a_{32} & a_{33} \end{bmatrix} \cdot \begin{bmatrix} x \\ y \\ z \end{bmatrix} \tag{5-28}$$

where the expressions containing the coordinates are called column vectors:

$$\mathbf{r} = \begin{bmatrix} x \\ y \\ z \end{bmatrix} \qquad\qquad \mathbf{r}' = \begin{bmatrix} x' \\ y' \\ z' \end{bmatrix} \tag{5-29}$$

and the square array of a_{ij} elements is a 3×3 square matrix.

By comparing Eqs. (5-28) and (5-1) one obtains

$$\mathbf{R} = \begin{bmatrix} a_{11} & a_{12} & a_{13} \\ a_{21} & a_{22} & a_{23} \\ a_{31} & a_{32} & a_{33} \end{bmatrix} = \{a_{ij}\} \tag{5-30}$$

Matrices satisfying conditions (5-27) are called orthogonal. Matrices which represent operations of symmetry are orthogonal.

The identity operator is

$$\mathbf{E} = \begin{bmatrix} 1 & 0 & 0 \\ 0 & 1 & 0 \\ 0 & 0 & 1 \end{bmatrix} \tag{5-31}$$

since $x' = x$, $y' = y$, and $z' = z$.

In order to find the multiplication procedure, we write two symmetry operators in matrix notation:

$$\mathbf{r'} = \{a_{ij}\} \cdot \mathbf{r} \tag{5-32}$$

$$\mathbf{r''} = [b_{ij}] \cdot \mathbf{r'} \tag{5-33}$$

equivalent to Eqs. (5-1) and (5-2). The symmetry operator of Eq. (5-4) may be expressed

$$\mathbf{r''} = [c_{ij}] \cdot \mathbf{r} \tag{5-34}$$

Eq. (5-5) then becomes

$$\{c_{ij}\} = [b_{ij}] \cdot \{a_{ij}\} \tag{5-35}$$

We obtain the elements of the matrix $[c_{ij}]$ by performing the operations indicated in Eq. (5-27), first by applying $[a_{ij}]$ to $\mathbf{r}(x, y, z)$, then by applying $\{b_{ij}\}$ to $\mathbf{r'}(x', y', z')$ to finally obtain $\mathbf{r''}(x'', y'', z'')$. Arrangement of variables shows that the coefficients c_{ij} of the matrix $\{c_{ij}\}$ which transforms $\mathbf{r}$ into $\mathbf{r''}$ are

$$c_{ji} = \sum_{k=1}^{3} (b_{jk} \cdot a_{ki}) \tag{5-36}$$

which is the well-known procedure of matrix multiplication.

The value of the terms a_{ij} of the matrix of Eq. (5-30) depends on the choice of coordinates. It can be verified that the reflection shown in Fig. 5.5 is expressed by two different matrices, according to the choice of coordinates:

$$\mathbf{m}(a) = \begin{bmatrix} \bar{1} & 0 & 0 \\ 0 & 1 & 0 \\ 0 & 0 & 1 \end{bmatrix} ; \mathbf{m}(b) = \begin{bmatrix} 0 & \bar{1} & 0 \\ \bar{1} & 0 & 0 \\ 0 & 0 & 1 \end{bmatrix} \tag{5-37}$$

In order to apply the matrix notation developed in this section, the symmetry of the water molecule should be studied. Fig. 5-7 illustrates a water molecule referred to a frame that provides the simplest matrices for each operation of symmetry (displayed at the bottom of the figure). The elements of symmetry are the digonal axis C_2 parallel to the z-axis and passing through the origin, and two orthogonal planes of symmetry, σ_v and σ_v', whose intersection coincides with the C_2 axis. The symmetry group of the water molecule is then $(\mathbf{E}, \mathbf{2}, \mathbf{m}, \mathbf{m'})$ where $\mathbf{2}$, $\mathbf{m}$, and $\mathbf{m'}$ are the symmetry operations generated by the elements C_2, σ_v, and σ_v'. The multiplication table of the group is shown in Table 5.3 and may be verified, as a useful exercise, by multiplying the matrices displayed in Fig. 5.7.

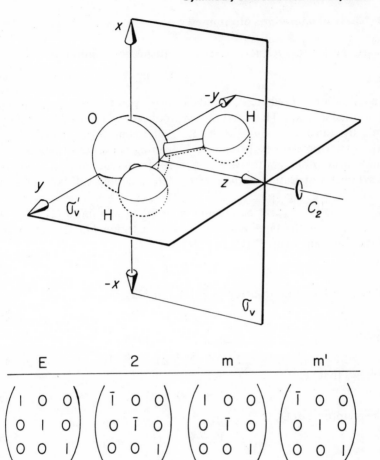

$$
E \qquad\qquad 2 \qquad\qquad m \qquad\qquad m'
$$

$$
\begin{pmatrix} 1 & 0 & 0 \\ 0 & 1 & 0 \\ 0 & 0 & 1 \end{pmatrix}
\begin{pmatrix} \bar{1} & 0 & 0 \\ 0 & \bar{1} & 0 \\ 0 & 0 & 1 \end{pmatrix}
\begin{pmatrix} 1 & 0 & 0 \\ 0 & \bar{1} & 0 \\ 0 & 0 & 1 \end{pmatrix}
\begin{pmatrix} \bar{1} & 0 & 0 \\ 0 & 1 & 0 \\ 0 & 0 & 1 \end{pmatrix}
$$

FIG. 5-7 Symmetry elements and operations of the water molecule.

TABLE 5.3 Table of multiplication of the water molecule symmetry group.

	E	2	m	m'
E	E	2	m	m'
2	2	E	m'	m
m	m	m'	E	2
m'	m'	m	2	E

5.9 Class of operations of symmetry

If a set of operators $\mathbf{R}_1, \mathbf{R}_2, \ldots, \mathbf{R}_i, \ldots$ satisfies the condition

$$\mathbf{R}_i = \mathbf{R}_g^{-1} \cdot \mathbf{R}_j \cdot \mathbf{R}_g \qquad (5\text{-}38)$$

with $i \neq j$ or $i = j$, where $\mathbf{R}_g$ stands for any operator of the group, such a set is called a *class*. The operation symbolized by Eq. (5-38) is called the similarity transformation of $\mathbf{R}_j$ by $\mathbf{R}_g$. The concept of class is very important in symmetry. Its geometrical meaning is the set of operators that transform into one another by a change of reference frame, which itself is a symmetry operation of the group. This meaning will become clear in the following paragraphs.

Let $\mathbf{r}_1 = P(x_1, y_1, z_1)$ be a point referred to a frame $F(\text{I})$. Let $\mathbf{r}_2 = P(x_2, y_2, z_2)$ be the same point referred to a frame $F(\text{II})$. Let $\mathbf{Q}$ be the matrix that transforms $F(\text{I})$ into $F(\text{II})$, i.e.,

$$\mathbf{r}_2 = \mathbf{Q} \cdot \mathbf{r}_1 \qquad (5\text{-}39)$$

and $\mathbf{S}$ the matrix that transforms $F(\text{II})$ into $F(\text{I})$, i.e.,

$$\mathbf{r}_1 = \mathbf{S} \cdot \mathbf{r}_2 \qquad (5\text{-}40)$$

Since

$$\mathbf{r}_1 = \mathbf{S} \cdot \mathbf{r}_2 = \mathbf{S} \cdot \mathbf{Q} \cdot \mathbf{r}_1 = \mathbf{E} \cdot \mathbf{r}_1 \qquad (5\text{-}41)$$

it follows that

$$\mathbf{S} = \mathbf{Q}^{-1} \qquad (5\text{-}42)$$

is the reciprocal of $\mathbf{Q}$, i.e.,

$$\mathbf{r}_1 = \mathbf{Q}^{-1} \cdot \mathbf{r}_2 \qquad (5\text{-}43)$$

Now let $\mathbf{r}_1'$ be an equivalent point referred to $F(\text{I})$ and $\mathbf{r}_2'$ the same point referred to $F(\text{II})$. There will be two expressions $\mathbf{R}_1$ and $\mathbf{R}_2$ for the symmetry operator that transforms $\mathbf{r}_i$ into $\mathbf{r}_i'$

$$\mathbf{r}_1' = \mathbf{R}_1 \cdot \mathbf{r}_1 \qquad (5\text{-}44)$$

and

$$\mathbf{r}_2' = \mathbf{R}_2 \cdot \mathbf{r}_2 \qquad (5\text{-}45)$$

If we replace $\mathbf{r}_2$ by its expression given in Eq. (5-39) we obtain

$$\mathbf{r}_2' = \mathbf{R}_2 \cdot \mathbf{Q} \cdot \mathbf{r}_1 \qquad (5\text{-}46)$$

If we operate Eq. (5-46) by $\mathbf{Q}^{-1}$ we get, in view of Eq. (5-43):

$$\mathbf{r}_1' = \mathbf{Q}^{-1} \cdot \mathbf{r}_2' = \mathbf{Q}^{-1} \cdot \mathbf{R}_2 \cdot \mathbf{Q} \cdot \mathbf{r}_1 \qquad (5\text{-}47)$$

If we now compare Eqs. (5-44) and (5-47) we find

$$\mathbf{R}_1 = \mathbf{Q}^{-1} \cdot \mathbf{R}_2 \cdot \mathbf{Q} \qquad (5\text{-}48)$$

which is the similarity transformation defined in Eq. (5-38).

Operating both terms of Eq. (5-48) by $\mathbf{Q}$ on the left and by $\mathbf{Q}^{-1}$ on the right it is found that

$$\mathbf{Q} \cdot \mathbf{R}_1 \cdot \mathbf{Q}^{-1} = \mathbf{Q} \cdot \mathbf{Q}^{-1} \cdot \mathbf{R}_2 \cdot \mathbf{Q}^{-1} \cdot \mathbf{Q} = \mathbf{E} \cdot \mathbf{R}_2 \cdot \mathbf{E} = \mathbf{R}_2 \qquad (5\text{-}49)$$

i.e.,

$$\mathbf{R}_2 = \mathbf{Q} \cdot \mathbf{R}_1 \cdot \mathbf{Q}^{-1} \qquad (5\text{-}50)$$

Since $\mathbf{r}_1' = P'(x_1', y_1', z_1')$ and $\mathbf{r}_2' = P'(x_2', y_2', z_2')$ are two different vectors representing the same point P', this latter point being equivalent to the point P, whose representations in both frames are $\mathbf{r}_1$ and $\mathbf{r}_2$, the operators $\mathbf{R}_1$ and $\mathbf{R}_2$ represent the same operation of symmetry referred to two different frames. The operation that transforms operators in a change of coordinates, indicated in Eqs. (5-48) and (5-50), is then the similarity transformation introduced in Eq. (5-38). The operator $\mathbf{R}_1$ is called the similarity transform of $\mathbf{R}_2$ by $\mathbf{Q}$. The operator $\mathbf{R}_2$, in turn, is the similarity transform of $\mathbf{R}_1$ by $\mathbf{Q}^{-1}$.

If $\mathbf{Q}$ is itself a symmetry operator of the group of $\mathbf{R}_1$ and $\mathbf{R}_2$, it will change any element of symmetry into an equivalent element of symmetry, and eventually into itself, since the element of symmetry is a geometrical entity. It then follows that if, for example, C_n and C_n' are two equivalent n-fold rotation axes generating the operations $\mathbf{R}$ and $\mathbf{R}'$, it can be stated that if

$$C_n' = \mathbf{Q} \cdot C_n \qquad (5\text{-}51)$$

then

$$\mathbf{R}' = \mathbf{Q} \cdot \mathbf{R} \cdot \mathbf{Q}^{-1} \qquad (5\text{-}52)$$

for Eq. (5-51) is equivalent to having changed the frame of reference of C_n.

The concept of class is important, for it permits a simplified representation of the symmetry group of a given configuration. Returning to the symmetry of an equilateral triangle, developed in Section 5.6, it can now be seen by observing the multiplication table (Table 5.2) that the operators $\mathbf{3}$ and $\mathbf{3}^2$ belong to a class, and that the three reflections belong to another class. Since $\mathbf{E}$ is a class in itself, we may represent the symmetry group of the equilateral triangle in the simplified way $(E, 2C_3, 3\sigma)$, which means: the class of one identity operation, the class of two 3-fold rotations, and the class of three reflections. That $\mathbf{3}$ and $\mathbf{3}^2$ form a class is evident from the two following considerations: (1) $\mathbf{3}^2$ is equivalent to $\mathbf{3}^{-1}$, a counterclockwise rotation by $2\pi/3$, and (2) any of the reflections transform a clockwise rotation about an axis perpendicular to the triangle into a counterclockwise rotation about the same axis. All the reflections belong to a class, resulting from the fact that the three planes of symmetry are exchanged among themselves by 3-fold rotations about an axis perpendicular to the triangle.

It is clear that if a set of operations of symmetry forms a class, only the influence of one of them on the properties of the configuration needs be studied, for such properties obviously cannot depend on a change of coordinates, which is a symmetry operation. In dealing with crystal and ligand field theory as applied to a regular octahedral environment which has forty-eight symmetry operations, for example, the existence of classes reduces the number of operations needed for solving the symmetry relations to only nine, plus the trivial operation of identity. This problem is examined in detail in Section 5.13.

5.10 The character of the symmetry matrix

In group theory, the sum of the diagonal terms of a matrix, otherwise called *trace*, is known as the *character* of the matrix. It is symbolized by

$$\chi = \sum_j a_{jj}. \tag{5-53}$$

An important property of a class of operations of symmetry is that the characters of the matrices representing operations within the class are equal, because *the character of a matrix is an invariant under the similarity transformation.* The proof follows.

Let **D** be the similarity transform of **C** by **B**, i.e.,

$$\mathbf{D} = \mathbf{B}^{-1} \cdot \mathbf{C} \cdot \mathbf{B} \tag{5-54}$$

Since

$$\mathbf{B} \cdot \mathbf{B}^{-1} = \mathbf{E}, \tag{5-55}$$

Eq. (5-54) becomes, after left-multiplication by **B**:

$$\mathbf{B} \cdot \mathbf{D} = \mathbf{C} \cdot \mathbf{B} \tag{5-56}$$

In general, we can find a matrix **A** such that

$$\mathbf{C} = \mathbf{B} \cdot \mathbf{A} \tag{5-57}$$

It then follows that

$$\mathbf{B} \cdot \mathbf{D} = \mathbf{B} \cdot \mathbf{A} \cdot \mathbf{B} \tag{5-58}$$

and hence

$$\mathbf{D} = \mathbf{A} \cdot \mathbf{B}. \tag{5-59}$$

Equations (5-57) and (5-59) are the conditions that conjugate matrices (**C** and **D**) satisfy. These equations may now be used to prove that the characters of **C** and **D** are equal, as follows.

Using the matrix multiplication procedure, one writes

$$= \sum_j c_{jj} = \sum_j \sum_k b_{jk} a_{kj} \qquad (5\text{-}60)$$

$$= \sum_k d_{kk} = \sum_k \sum_j a_{kj} b_{jk}. \qquad (5\text{-}61)$$

Since the elements a and b are numbers,

$$b_{jk} a_{kj} = a_{kj} b_{jk}, \qquad (5\text{-}62)$$

and one arrives at

$$\chi_C = \chi_D, \qquad (5\text{-}63)$$

Q.E.D.

This property is very important in the representation of groups, as it will be seen when treating irreducible representations and character tables, for if the characters are taken for the representation, this latter will be independent of the reference frame.

5.11 Notation

There are two broad fields in which considerations of symmetry are of paramount importance: *crystallography* and *molecular spectroscopies*. We will rapidly review the notation used. The main difference between both fields lies in the improper axis S_n of molecular spectroscopists and the inversion axis of crystallographers. In crystallography, an inversion axis is an element about which an n-fold rotation followed by an inversion is performed. The international symbol is $\bar{n}$. In crystals only 2-, 3-, 4-, and 6-fold axes appear; they generate the operations $\bar{2}, \bar{3}, (\bar{3})^2, \dots, (\bar{6})^5$. The inversion 2-fold axis is equivalent to a plane of symmetry perpendicular to the axis.

Proof: If the $\bar{2}$-fold axis is brought into coincidence with the z-axis, then

$$\bar{2}_z = \bar{1} \cdot 2 = \begin{bmatrix} \bar{1} & 0 & 0 \\ 0 & \bar{1} & 0 \\ 0 & 0 & \bar{1} \end{bmatrix} \cdot \begin{bmatrix} \bar{1} & 0 & 0 \\ 0 & \bar{1} & 0 \\ 0 & 0 & 1 \end{bmatrix} \qquad (5\text{-}64)$$

Making the matrix product, one finds

$$\bar{2}_z = \begin{bmatrix} 1 & 0 & 0 \\ 0 & 1 & 0 \\ 0 & 0 & \bar{1} \end{bmatrix} = \mathbf{m}_{xy} \tag{5-65}$$

which is the matrix that represents a reflection in the plane (x, y), i.e., that changes z into $-z$ and leaves x and y unchanged.

In molecular spectroscopy, an improper axis S_n is an element about which an n-fold rotation followed by a reflection about a plane perpendicular to the improper axis is performed. The operation S_2 is equivalent to an inversion since, again using the z-axis coincident with S_2, it can be written

$$S_2 = \mathbf{m}_h \cdot 2 = \begin{bmatrix} 1 & 0 & 0 \\ 0 & 1 & 0 \\ 0 & 0 & \bar{1} \end{bmatrix} \cdot \begin{bmatrix} \bar{1} & 0 & 0 \\ 0 & \bar{1} & 0 \\ 0 & 0 & 1 \end{bmatrix} \tag{5-66}$$

and the matrix product gives the inversion operator

$$\bar{1} = \begin{bmatrix} \bar{1} & 0 & 0 \\ 0 & \bar{1} & 0 \\ 0 & 0 & \bar{1} \end{bmatrix} = S_2 \tag{5-67}$$

The reader may verify, as a useful exercise, the following equalities:

$$(\bar{6})^3 = \mathbf{m}; \quad S_6^3 = \bar{1} \tag{5-68}$$

TABLE 5.4   Notation of symmetry operations.

Operation	Symbol Schoenflies	Symbol International
reflection	σ	$\mathbf{m}$
n-fold proper rotation	C_n	$n = 2, 3, 3^2, \ldots$
n-fold improper rotation	S_n	
n-fold inversion rotation		$\bar{2} = \mathbf{m}, \bar{3}, \ldots$
n-fold screw rotation		$2_1, 3_1, \ldots$
inversion	i	$\bar{1}$
glide planes		a, b, c, n, d
identity	E	1

Since considerations involving the crystallographic screw axes are out of the scope of this book, they will not be discussed. If needed, we will use the crystallographic notation n_m for the operation, m being the pitch. Table 5.4 displays the symbols used in this book and their meaning.

5.12 Molecular symmetry and symmetry point groups

The groups of operators that describe the symmetry of molecules are called *point groups*, the reason being that their operations leave at least one point of the configuration unchanged. There are also *space groups*, which involve translations and are of interest in crystallography. These latter are not considered in this book.

The symmetry point groups are designated by a notation originally developed by Schoenflies, and incorporated in the specialized literature under his name. The element notations E, C_n, S_n, i, and σ have already been introduced in preceding sections. Before developing Schoenflies' notation of the point groups, however, it is necessary to introduce a more detailed notation of the symmetry planes, which are further classified as follows.

If the plane is perpendicular to the principal axis of symmetry (the n-fold proper or improper axis of highest order n), it is symbolized σ_h. If the plane contains the principal axis, it is symbolized σ_v. If there are 2-fold axes perpendicular to the principal symmetry axis and planes of symmetry containing the principal axis and bisecting the angle between 2-fold axes, these planes are symbolized σ_d. The molecular symmetry groups may then be classified as shown by the following outline.

I. The Rotation Groups
   A. The molecule possesses only one n-fold axis. The groups of interest are: C_1, C_2, C_3, C_4, C_5, C_6, C_7, and (C_8).
   B. The molecule possesses one n-fold axis and σ_v. The groups of interest are: C_{2v}, C_{3v}, C_{4v}, C_{5v}, C_{6v}, ...
   C. The molecule possesses one n-fold axis and σ_h. The groups of interest are: $C_{1h}(=C_s)$, C_{2h}, C_{3h}, C_{4h}, C_{5h}, ...
   D. The molecule has an improper axis S_n. The groups are: S_2 $(=C_i)$, S_4, S_6 $(=C_{3i})$, ... The equivalence $S_2 = C_i$ was shown in Eq. (5-56.) The equivalence $S_6 = C_{3i}$ follows.

Notice that only even values of n give rise to improper rotation groups. This is due to the fact that when n is odd

$$S_n^m = n^m \qquad (5-69)$$

for m even, while for both n and m odd we may write

$$S_n^m = n^m \cdot m_h \tag{5-70}$$

Therefore S_n with n odd is equivalent to C_{nh}. In other words, should n be odd, performing n times the n-fold improper rotation would leave the configuration reflected about the plane σ_h; hence, σ_h must be an element of symmetry. Performing $(n - 1)$ times such improper rotation would be equivalent to $(n - 1)$ times the n-fold proper rotation, for $(n - 1)$ would now be even. Since there exists the element σ_h, there must exist the element C_n (n = odd), and the group is C_{nh} (n = odd).

E.   The molecule has, in addition to the principal n-fold axis, $2n$ 2-fold axes perpendicular to it. These axes are denoted by C_2, C_2', C_2'', ... The notation for these groups is D_n. Possible groups are: $D_2(=V)$, D_3, D_4, D_5, D_6, ...

F.   The molecule possesses the elements of D_n plus σ_d. The groups of interest are: $D_{2d}(=V_d)$, D_{3d}, D_{4d}, D_{5d}, ...

G.   The molecule has, in addition to the elements of the group D_{nd}, the element σ_h. The possible groups are: $D_{2h}(=V_h)$, D_{3h}, D_{4h}, D_{5h}, D_{6h}, ...

II.   Point Groups of Higher Symmetry

These are the groups that have more than one n-fold axis of maximum n; the most important are:

T_d: the point group of the regular tetrahedron
O_h: the point group of the regular octahedron
I_h: the point group of the regular icosahedron and dodecahedron

III.   Point Groups of Linear Molecules

There are two point groups, according to whether or not the element σ_h exists (i.e., $D_{\infty h}$ and $C_{\infty v}$).

Examples of molecules and ions belonging to various symmetry groups are displayed in Table 5.5.

5.13   Cubic, octahedral, and tetrahedral symmetries

The highest-symmetry tridimensional shapes are found in the cubic system. We will restrict the study to three configurations often found in chemistry: cube, regular octahedron, and regular tetrahedron. The cube and the regular octahedron have 48 operations of symmetry, the regular tetrahedron, 24. Since these configurations are of special interest in crystal and ligand-field theories, as well as in the construction of molecular

TABLE 5.5 Examples of molecules and ions belonging to various symmetry groups.

Point group	Molecules
C_s	BFClBr
C_{2v}	H_2O SO_2 H_2S
C_{3v}	NH_3 PCl_3 CH_3Cl
C_{2h}	*trans*-$C_2H_2Cl_2$
D_{3h}	PCl_5 (trigonal bipyramid)
D_{4h}	$(AuCl_4)^-$
D_{5h}	$(C_5H_5)_2Ru$ (ruthenocene)
D_{6h}	C_6H_6
D_{2d}	$H_2C{=}C{=}CH_2$ (allene)
D_{5d}	$(C_5H_5)_2Fe$ (ferrocene)
$C_{\infty v}$	CO HCl
$D_{\infty h}$	H_2 O_2 C_2H_4
T_d	CH_4
O_h	SF_6
I_h	$(B_{12}H_{12})^{2-}$

orbitals by linear combination of atomic orbitals (LCAO), their symmetry operations will be discussed in some detail.

Cube and Octahedron.   The elements and operations of symmetry are:

(1) Four S_6 axes along the directions normal to the planes of Miller indices (111), $(1\bar{1}1)$, $(\bar{1}\bar{1}1)$, and $(11\bar{1})$, each of them generating the operations $\bar{3}, \bar{3}^2 = 3^2, \bar{3}^3 = \bar{1}, \bar{3}^4 = 3$, and $\bar{3}^5 = \bar{3}^{-1}$.

(2) Four C_3 axes along the S_6 directions, generating the operations **3** and 3^2, already counted in (1).

(3) A center of symmetry, *i*, generating the operation $\bar{1}$, already counted in (1).

(4) Three S_4 axes normal to the planes (100), (010), and (001), generating the operations $\bar{4}, \bar{4}^2 = 2$, and $\bar{4}^3 = \bar{4}^{-1}$.

(5) Three C_4 axes along the directions of S_4, generating the operations **4**, $4^2 = 2$, and $4^3 = 4^{-1}$. Among them, **4** and $4^3 = 4^{-1}$ are new.

(6) Three C_2 axes along the directions of S_4, generating the operations **2** $(= 4^2 = \bar{4}^2)$ already counted.

(7) Six C_2' axes normal to the planes (110), (101), (011), $(\bar{1}10)$, $(\bar{1}01)$, and $(0\bar{1}1)$ respectively, each one generating an operation **2'**.

(8) Six σ of Miller indices (110), (101), (011), $(\bar{1}10)$, $(\bar{1}01)$, and $(0\bar{1}1)$, each generating an operation **m**.

(9)   Three σ' of Miller indices (100), (010), and (001), each generating an operation **m'**.

(10)   The identity operation **E**.

The total number of operations is 48. They are divided in classes as follows. Operations $\bar{3}$ and $\bar{3}^5$ belong to a class that contains eight operations of symmetry about S_6. We call this class $8S_6$. Operations 3 and 3^2 belong to the class $8C_3$. Operation $\bar{1}$ is a class itself: *i*. There is, in addition, the identity class *E*. The remaining operations belong to the classes $6S_4$, $6C_4$, $3C_2(4^2)$, $6C'_2$, 6σ, and $3\sigma'$.

As a useful application of the concept of class developed in Section 5-10, we will prove that the operations 3 and 3^2 form a class. We first write the operations in matrix notation by adopting a frame of reference with the origin at the center of symmetry of the cube and the reference axes parallel to the cube edges. Fig. 5.8 illustrates the position of the ref-

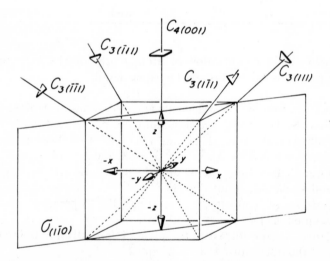

FIG. 5-8   Some elements of symmetry of a cube.

erence frame as well as that of the elements of symmetry under consideration. The 3-fold operators are:

$$3(111) = \begin{bmatrix} 0 & 0 & \bar{1} \\ 1 & 0 & 0 \\ 0 & 1 & 0 \end{bmatrix} \qquad 3^2(111) = \begin{bmatrix} 0 & 1 & 0 \\ 0 & 0 & 1 \\ 1 & 0 & 0 \end{bmatrix} \qquad (5\text{-}71)$$

$$3(1\bar{1}1) = \begin{bmatrix} 0 & \bar{1} & 0 \\ 0 & 0 & \bar{1} \\ 1 & 0 & 0 \end{bmatrix} \qquad 3^2(1\bar{1}1) = \begin{bmatrix} 0 & 0 & 1 \\ \bar{1} & 0 & 0 \\ 0 & \bar{1} & 0 \end{bmatrix} \tag{5-72}$$

$$3(\bar{1}\bar{1}1) = \begin{bmatrix} 0 & 0 & \bar{1} \\ 1 & 0 & 0 \\ 0 & \bar{1} & 0 \end{bmatrix} \qquad 3^2(\bar{1}\bar{1}1) = \begin{bmatrix} 0 & 1 & 0 \\ 0 & 0 & \bar{1} \\ \bar{1} & 0 & 0 \end{bmatrix} \tag{5-73}$$

$$3(\bar{1}11) = \begin{bmatrix} 0 & 0 & \bar{1} \\ \bar{1} & 0 & 0 \\ 0 & 1 & 0 \end{bmatrix} \qquad 3^2(\bar{1}11) = \begin{bmatrix} 0 & \bar{1} & 0 \\ 0 & 0 & 1 \\ \bar{1} & 0 & 0 \end{bmatrix} \tag{5-74}$$

Since the **3** and 3^2 operators generated by each C_3 element are reciprocal, one may write, given the matrix of one of them, the matrix of the other by exchanging columns and rows. Because the matrices are real orthogonal, this is a general property of reciprocal operators.

In order to prove that the operators **3** belong to a class, we must search for a similarity transformation by an operator of symmetry that transforms the elements C_3 into one another. One such operator is **4**(001). Its matrix notation is

$$4(001) = \begin{bmatrix} 0 & \bar{1} & 0 \\ 1 & 0 & 0 \\ 0 & 0 & 1 \end{bmatrix} \tag{5-75}$$

and that of its reciprocal (by exchanging columns and rows):

$$4^{-1}(001) = 4^3(001) = \begin{bmatrix} 0 & 1 & 0 \\ \bar{1} & 0 & 0 \\ 0 & 0 & 1 \end{bmatrix} \tag{5-76}$$

The symmetry elements transform according to

$$C_3(1\bar{1}1) = 4(001) \cdot C_3(111) \tag{5-77}$$

$$C_3(\bar{1}\bar{1}1) = 4(001) \cdot C_3(1\bar{1}1) \tag{5-78}$$

$$C_3(\bar{1}11) = 4(001) \cdot C_3(\bar{1}\bar{1}1) \tag{5-79}$$

$$C_3(111) = 4(001) \cdot C_3(\bar{1}11) \tag{5-80}$$

These operations may be performed by writing the elements as column vectors, i.e.,

$$\begin{bmatrix} 1 \\ \bar{1} \\ 1 \end{bmatrix} = \begin{bmatrix} 0 & \bar{1} & 0 \\ 1 & 0 & 0 \\ 0 & 0 & 1 \end{bmatrix} \cdot \begin{bmatrix} 1 \\ 1 \\ 1 \end{bmatrix} \tag{5-81}$$

and so on. Notice that it is immaterial to distinguish between $C_3(111)$ and $C_3(\bar{1}\bar{1}\bar{1})$, $C_3(1\bar{1}1)$ and $C_3(\bar{1}1\bar{1})$, etc., for the elements generating rotations are axes which are completely defined by the direction and not the sense. Hence, they are centrosymmetrical.

The similarity transformations are:

$$3(1\bar{1}1) = 4(001) \cdot 3(111) \cdot 4^{-1}(001) \tag{5-82}$$

$$3(\bar{1}\bar{1}1) = 4(001) \cdot 3(1\bar{1}1) \cdot 4^{-1}(001) \tag{5-83}$$

$$3(\bar{1}11) = 4(001) \cdot 3(\bar{1}\bar{1}1) \cdot 4^{-1}(001) \tag{5-84}$$

$$3(111) = 4(001) \cdot 3(\bar{1}11) \cdot 4^{-1}(001) \tag{5-85}$$

Similar equations are obtained relating the 3^2 operators. Now it is necessary to prove that any operator 3^2 is the similarity transform of an operator 3 by a symmetry operator. Since $3^2 = 3^{-1}$, such a symmetry operator must transform a clockwise into a counterclockwise rotation about, for example, the element $C_3(111)$, and leave the element unchanged. An operation fulfilling these requirements is the reflection about the symmetry plane $\sigma(1\bar{1}0)$, whose matrix representation is

$$\mathbf{m}(1\bar{1}0) = \begin{bmatrix} 0 & 1 & 0 \\ 1 & 0 & 0 \\ 0 & 0 & 1 \end{bmatrix} \tag{5-86}$$

We then have

$$3^2(111) = \mathbf{m}(110) \cdot 3(111) \cdot \mathbf{m}(110) \tag{5-87}$$

since $\mathbf{m}(110)$ is its own reciprocal. It has thus been proven that the four operations 3 and the four operations 3^2 form a class of operators generated by the elements C_3, therefore called $8C_3$.

Let us point out that the operators 4 and 4^3 belong to a class, but that the operator 4^2, generated by the same element of symmetry C_4, does not belong in the same class. This fact has a simple geometrical explanation: there is no change of reference frame that transforms a rotation by $1/2\pi$ into a rotation by π. Analogously, in configurations having the element C_3 but not the symmetry plane that transforms a 3-fold clockwise into a 3-fold counterclockwise rotation, the operators 3 and 3^2 do not belong in the same class.

Returning to the subject of this section, a representation by classes of the cube and regular octahedron is then: E, i, $8S_6$, $8C_3$, $6S_4$, $6C_4$, $3C_2$, $6C'_2$, $3\sigma_h$, 6σ. The reader may verify that a representation by classes of the regular tetrahedron symmetry group $\mathbf{T}_d$ is (E, $8C_3$, $6S_4$, $3C_2$, 6σ), adding up to 24 symmetry operations.

5.14   Systems of lower symmetry

The cubic forms studied in Section 5.13 may be distorted in two ways: by changing the height of the parallelepiped (i.e., by distorting the cube along the z-axis (001)), or by distorting the figure along one of the trigonal axes (111). The first case leads to the tetragonal system, the second to the rhombohedral and hexagonal systems. Further distortion of the tetragonal lattice along the y-axis destroys the 4-fold axis, giving rise to the

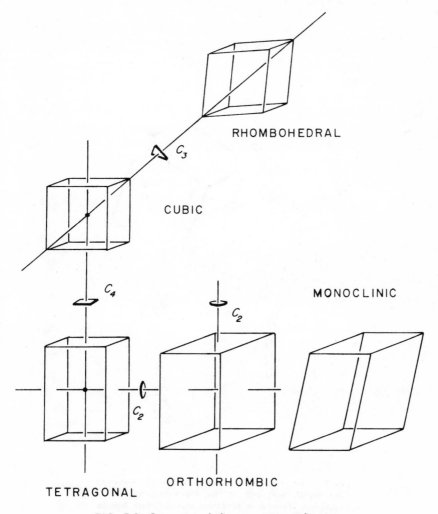

FIG. 5-9   Symmetry relations among crystal systems.

orthorhombic lattice, still characterized by three orthogonal unit-cell vectors. When one of the angles between two unit-cell vectors is changed, a monoclinic lattice is obtained. If the other two are now distorted, one arrives to the less symmetric system: triclinic. In order to keep our discussion within the limits imposed by the scope of the book, we need not consider systems of lesser symmetry than orthorhombic. The distortion along the z-axis of the cube that generates the tetragonal system removes the 3-fold axes and two of the three 4-fold axes, leaving only one proper or

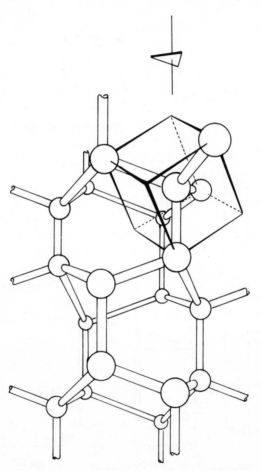

FIG. 5-10 $P6_2/mmc$ lattice of ordinary ice showing one-eighth unit cell of the diamond-cube, $F\bar{4}3m$ structure. (From J. A. McMillan and S. C. Los, Nature **206**, 806 (1965), by permission of the publisher.)

improper 4-fold axis, along z. Of particular interest among the tetragonal groups is the symmetry group of the square planar configuration, often found in complex ions. Its Schoenflies' point group is $\mathbf{D}_{4h}$. It has 16 symmetry operations distributed among 10 classes, namely: E, $2C_4$, C_2, $2C_2'$, $2C_2''$, i, $2S_4$, σ_h, $2\sigma_v$, $2\sigma_d$. If one assumes that the four ligand molecules may be treated as point charges, the group reduces to $\mathbf{C}_{4v}$ (E, $2C_4$, C_2, $2\sigma_v$, $2\sigma_d$). Groups of tetragonal symmetry have operators that exchange x and y, but not z. Analogously, rhombohedral and hexagonal symmetry groups mix x and y, but not z. These cases are discussed in Chapter 7.

The distortion of the tetragonal system that generates the orthorhombic lattices removes the equivalence of x and y by removing the 4-fold axis. None of the remaining symmetry operations mixes x, y, or z. The property of mixing or not mixing x, y, and z is of fundamental importance in the removal of orbital degeneracy. This situation will be examined in Chapter 12.

Fig. 5.9 shows the relationship among several systems and Fig. 5.10 the relationship between the diamond cube space group $\mathbf{F\bar{4}3m}$ and the hexagonal space group $\mathbf{P6_3/mmc}$.*

5.15 Representation of groups

We have seen that the multiplication table of a group is a representation of the group. In applying symmetry operations to an equilateral triangle we were able to construct a 6×6 table of symmetry operations giving *the rules of the game* for the particular group under consideration. If we now take an array of three elements (α, β, γ) without any specific meaning, and make all the permutations, we may name each operation by a capital letter, as follows.

$$\left.\begin{array}{lll} \mathbf{E} = \begin{pmatrix} \alpha & \beta & \gamma \\ \alpha & \beta & \gamma \end{pmatrix} & \mathbf{A} = \begin{pmatrix} \alpha & \beta & \gamma \\ \alpha & \gamma & \beta \end{pmatrix} & \mathbf{B} = \begin{pmatrix} \alpha & \beta & \gamma \\ \gamma & \beta & \alpha \end{pmatrix} \\[1.5em] \mathbf{C} = \begin{pmatrix} \alpha & \beta & \gamma \\ \beta & \alpha & \gamma \end{pmatrix} & \mathbf{D} = \begin{pmatrix} \alpha & \beta & \gamma \\ \gamma & \alpha & \beta \end{pmatrix} & \mathbf{F} = \begin{pmatrix} \alpha & \beta & \gamma \\ \beta & \gamma & \alpha \end{pmatrix} \end{array}\right\} \quad (5\text{-}88)$$

where the original arrangement is shown at the top of each expression and the *operated* arrangement is shown at the bottom. It is easy to verify that the set ($\mathbf{E}$, $\mathbf{A}$, $\mathbf{B}$, $\mathbf{C}$, $\mathbf{D}$, $\mathbf{F}$) is a group, and that the elements of the group

*The symbols $\mathbf{F\bar{4}3m}$ and $\mathbf{P6_3}/mmc$ are international symbols for the crystallographic space groups. There are 230 space groups that result from the combination of the 32 crystallographic point groups with screw axes and glide mirrors.

form three classes: **E**, (**A**, **B**, **C**), and (**D**, **F**). The multiplication table is displayed in Table 5.6.

TABLE 5.6   Multiplication table of the group of permutations of three elements (α, β, γ).

	E	A	B	C	D	F
E	E	A	B	C	D	F
A	A	E	D	F	B	C
B	B	F	E	D	C	A
C	C	D	F	E	A	B
D	D	C	A	B	F	E
F	F	B	C	A	E	D

If we now compare the multiplication table of this group with the multiplication table of the equilateral triangle, we find that they are alike if we establish the one-to-one correspondence:

$$\left.\begin{array}{lll} \mathbf{E} \leftrightarrow \mathbf{E} & \mathbf{A} \leftrightarrow \sigma_A & \mathbf{B} \leftrightarrow \sigma_B \\ \mathbf{C} \leftrightarrow \sigma_C & \mathbf{D} \leftrightarrow C_3 & \mathbf{F} \leftrightarrow C_3^2 \end{array}\right\}. \qquad (5\text{-}89)$$

In this example we see that a group of symmetry transformations applied to a geometrical shape, and a group of permutations applied to an abstract set of elements have the same multiplication table. These groups are called *isomorphic;* there is no difference between them outside the specific meaning of each element of the group, for there is one-to-one correspondence between elements of each group. We may then expect representations derived without resorting to symmetry arguments, and this is what actually happened before group theoretical procedures made their way into the physical sciences. For the purpose of this book, however, we will resort to geometrical arguments and show, rather than prove, how to make representations of the groups of symmetry in the understanding that these representations can be mathematically founded in the most rigorous manner. The mathematically inclined reader will find the proofs in many of the books included in the list of suggested references.

The representation of a symmetry group is based on the manner in which certain vectors characterizing the configuration change upon application of the symmetry operators. Thus, if we refer the equilateral triangle to a tridimensional system of coordinates (x, y, z) with the origin in the center of figure, the x-axis parallel to the B-C side, and the z-axis perpendicular to the plane of the triangle, the operations of symmetry become

$$
\mathbf{E} = \begin{pmatrix} 1 & 0 & 0 \\ 0 & 1 & 0 \\ 0 & 0 & 1 \end{pmatrix}, \qquad
\sigma_A = \begin{pmatrix} -1 & 0 & 0 \\ 0 & 1 & 0 \\ 0 & 0 & 1 \end{pmatrix}
$$

$$
\sigma_B = \begin{pmatrix} \dfrac{1}{2} & \dfrac{\sqrt{3}}{2} & 0 \\ \dfrac{\sqrt{3}}{2} & -\dfrac{1}{2} & 0 \\ 0 & 0 & 1 \end{pmatrix}, \qquad
\sigma_C = \begin{pmatrix} \dfrac{1}{2} & -\dfrac{\sqrt{3}}{2} & 0 \\ -\dfrac{\sqrt{3}}{2} & -\dfrac{1}{2} & 0 \\ 0 & 0 & 1 \end{pmatrix}
$$

$$
\mathbf{C}_3 = \begin{pmatrix} -\dfrac{1}{2} & -\dfrac{\sqrt{3}}{2} & 0 \\ \dfrac{\sqrt{3}}{2} & -\dfrac{1}{2} & 0 \\ 0 & 0 & 1 \end{pmatrix}, \qquad
\mathbf{C}_3^2 = \begin{pmatrix} -\dfrac{1}{2} & \dfrac{\sqrt{3}}{2} & 0 \\ -\dfrac{\sqrt{3}}{2} & -\dfrac{1}{2} & 0 \\ 0 & 0 & 1 \end{pmatrix}
\tag{5-90}
$$

These matrices multiply according to the multiplication table and therefore are a representation of the group using the reference frame above described as a *basis for the representation.* If we choose an orthogonal set of coordinates such that the C_3-axis makes the same angle with all three coordinates, the representation is

$$
\mathbf{E} = \begin{pmatrix} 1 & 0 & 0 \\ 0 & 1 & 0 \\ 0 & 0 & 1 \end{pmatrix}, \qquad
\sigma_A = \begin{pmatrix} 1 & 0 & 0 \\ 0 & 0 & 1 \\ 0 & 1 & 0 \end{pmatrix},
$$

$$
\sigma_B = \begin{pmatrix} 0 & 0 & 1 \\ 0 & 1 & 0 \\ 1 & 0 & 0 \end{pmatrix}, \qquad
\sigma_C = \begin{pmatrix} 0 & 1 & 0 \\ 1 & 0 & 0 \\ 0 & 0 & 1 \end{pmatrix},
\tag{5-91}
$$

$$
\mathbf{C}_3 = \begin{pmatrix} 0 & 1 & 0 \\ 0 & 0 & 1 \\ 1 & 0 & 0 \end{pmatrix}, \qquad
\mathbf{C}_3^2 = \begin{pmatrix} 0 & 0 & 1 \\ 1 & 0 & 0 \\ 0 & 1 & 0 \end{pmatrix}
$$

Notice that the same set of matrices would be obtained if three coplanar vectors with origin in the center of figure and pointing toward each vertex (r_A, r_B, r_C) were used as a basis. The character of each class is, in both representations, the same, i.e.,

$$
\chi_E = 3, \quad \chi_{2C_3} = 0, \quad \chi_{3\sigma} = 1.
\tag{5-92}
$$

This is because in changing bases for the representation we have changed the reference frame and, since the operators are then related by a similarity transformation, their characters are invariant.

However, if we examine the first set of matrices, displayed in Eqs. (5-90), we see that in all cases z remains unchanged. The matrices are all *blocked out*; they are reducible to a lower order, indicated in dotted lines. The minimum order to which the set of matrices can be blocked out is clearly *two*, and the *reduced set* is now

$$
\left.
\begin{array}{lll}
\mathbf{E} = \begin{pmatrix} 1 & 0 \\ 0 & 1 \end{pmatrix}, & \mathbf{C}_3 = \begin{pmatrix} -\dfrac{1}{2} & -\dfrac{\sqrt{3}}{2} \\ \dfrac{\sqrt{3}}{2} & -\dfrac{1}{2} \end{pmatrix}, & \mathbf{C}_3^2 = \begin{pmatrix} -\dfrac{1}{2} & \dfrac{\sqrt{3}}{2} \\ -\dfrac{\sqrt{3}}{2} & -\dfrac{1}{2} \end{pmatrix} \\[3em]
\sigma_A = \begin{pmatrix} -1 & 0 \\ 0 & 1 \end{pmatrix}, & \sigma_B = \begin{pmatrix} \dfrac{1}{2} & \dfrac{\sqrt{3}}{2} \\ \dfrac{\sqrt{3}}{2} & -\dfrac{1}{2} \end{pmatrix} & \sigma_C = \begin{pmatrix} \dfrac{1}{2} & -\dfrac{\sqrt{3}}{2} \\ -\dfrac{\sqrt{3}}{2} & -\dfrac{1}{2} \end{pmatrix}
\end{array}
\right\} \quad (5\text{-}93)
$$

The characters are now

$$\chi_E = 2, \qquad \chi_{2C_3} = -1, \qquad \chi_{3\sigma} = 0 \qquad (5\text{-}94)$$

which are obtained by subtracting $+1$ from the characters of Eqs. (5-92). We will see later that

$$\chi_E = 1, \qquad \chi_{2C_3} = 1, \qquad \chi_{3\sigma} = 1 \qquad (5\text{-}95)$$

is also a representation of this group.

It is clear that the number of representations is only limited by our ability to imagine new and different bases. Another instructional example, the water molecule, will help us to fix these ideas and get more information about group representation.

Figure 5-7 shows the water molecule and its elements of symmetry, namely: ($\mathbf{E}$, $\mathbf{C}_2$, σ, σ'). In the reference frame indicated in the figure, the matrix representation of the group is

$$
\left.
\begin{array}{ll}
\mathbf{E} = \begin{pmatrix} 1 & 0 & 0 \\ 0 & 1 & 0 \\ 0 & 0 & 1 \end{pmatrix}, & \mathbf{C}_2 = \begin{pmatrix} -1 & 0 & 0 \\ 0 & -1 & 0 \\ 0 & 0 & 1 \end{pmatrix} \\[3em]
\sigma = \begin{pmatrix} 1 & 0 & 0 \\ 0 & -1 & 0 \\ 0 & 0 & 1 \end{pmatrix}, & \sigma' = \begin{pmatrix} -1 & 0 & 0 \\ 0 & 1 & 0 \\ 0 & 0 & 1 \end{pmatrix}
\end{array}
\right\} \quad (5\text{-}96)
$$

We write now the explicit operations using the matrices of Eqs. (5-96). Notice that the basis of this representation is the set of coordinates (x, y, z).

$$\mathbf{E}: \begin{pmatrix} x \\ y \\ z \end{pmatrix} = \begin{pmatrix} 1 & 0 & 0 \\ 0 & 1 & 0 \\ 0 & 0 & 1 \end{pmatrix} \cdot \begin{pmatrix} x \\ y \\ z \end{pmatrix} \tag{5-97}$$

$$\mathbf{C}_2: \begin{pmatrix} -x \\ -y \\ z \end{pmatrix} = \begin{pmatrix} -1 & 0 & 0 \\ 0 & -1 & 0 \\ 0 & 0 & 1 \end{pmatrix} \cdot \begin{pmatrix} x \\ y \\ z \end{pmatrix} \tag{5-98}$$

$$\sigma: \begin{pmatrix} x \\ -y \\ z \end{pmatrix} = \begin{pmatrix} 1 & 0 & 0 \\ 0 & -1 & 0 \\ 0 & 0 & 1 \end{pmatrix} \cdot \begin{pmatrix} x \\ y \\ z \end{pmatrix} \tag{5-99}$$

$$\sigma': \begin{pmatrix} -x \\ y \\ z \end{pmatrix} = \begin{pmatrix} -1 & 0 & 0 \\ 0 & 1 & 0 \\ 0 & 0 & 1 \end{pmatrix} \cdot \begin{pmatrix} x \\ y \\ z \end{pmatrix} \tag{5-100}$$

which are all blocked out so that the effect of each operator on each co-ordinate may be expressed as

E	C_2	σ	σ'	
1	-1	1	-1	x
1	-1	-1	1	y
1	1	1	1	z

Each row is now the representation of the group taking as basis x, y, z, respectively.

The complete description of a molecule must include rotations. We then chose rotations about the axes, R_x, R_y, R_z as bases. They will transform as

$$\mathbf{E}: \begin{pmatrix} R_x \\ R_y \\ R_z \end{pmatrix} = \begin{pmatrix} 1 & 0 & 0 \\ 0 & 1 & 0 \\ 0 & 0 & 1 \end{pmatrix} \cdot \begin{pmatrix} R_x \\ R_y \\ R_z \end{pmatrix} \tag{5-101}$$

$$\mathbf{C}_2: \begin{pmatrix} -R_x \\ -R_y \\ R_z \end{pmatrix} = \begin{pmatrix} -1 & 0 & 0 \\ 0 & -1 & 0 \\ 0 & 0 & 1 \end{pmatrix} \cdot \begin{pmatrix} R_x \\ R_y \\ R_z \end{pmatrix} \tag{5-102}$$

$$\sigma: \begin{pmatrix} -R_x \\ R_y \\ -R_z \end{pmatrix} = \begin{pmatrix} -1 & 0 & 0 \\ 0 & 1 & 0 \\ 0 & 0 & -1 \end{pmatrix} \cdot \begin{pmatrix} R_x \\ R_y \\ R_z \end{pmatrix} \qquad (5\text{-}103)$$

$$\sigma': \begin{pmatrix} R_x \\ -R_y \\ -R_z \end{pmatrix} = \begin{pmatrix} 1 & 0 & 0 \\ 0 & -1 & 0 \\ 0 & 0 & -1 \end{pmatrix} \cdot \begin{pmatrix} R_x \\ R_y \\ R_z \end{pmatrix} \qquad (5\text{-}104)$$

The change in sign of a rotation means that, say, a counterclockwise rotation is transformed into a clockwise rotation by the same angle. The effect of the application of each operator to each fundamental rotation is

E	C_2	σ	σ'	
1	-1	-1	1	R_x
1	-1	1	-1	R_y
1	1	-1	-1	R_z

If we compare the two foregoing tables of representations, we find that some of the representations are common. We then may construct the more complete table shown in Table 5.7, where A_1, A_2, B_1, B_2 are used, for the time being, merely to label each row, which is *an irreducible representation of the group* on the bases displayed in the third and fourth columns. Let us see how the quadratic-coordinate bases generate representations, by writing

$$E: \begin{pmatrix} x^2 \\ y^2 \\ z^2 \end{pmatrix} = \begin{pmatrix} 1 & 0 & 0 \\ 0 & 1 & 0 \\ 0 & 0 & 1 \end{pmatrix} \cdot \begin{pmatrix} x^2 \\ y^2 \\ z^2 \end{pmatrix},$$

$$\begin{pmatrix} xy \\ yz \\ xz \end{pmatrix} = \begin{pmatrix} 1 & 0 & 0 \\ 0 & 1 & 0 \\ 0 & 0 & 1 \end{pmatrix} \cdot \begin{pmatrix} xy \\ yz \\ xz \end{pmatrix} \qquad (5\text{-}105)$$

$$C_2: \begin{pmatrix} x^2 \\ y^2 \\ z^2 \end{pmatrix} = \begin{pmatrix} 1 & 0 & 0 \\ 0 & 1 & 0 \\ 0 & 0 & 1 \end{pmatrix} \cdot \begin{pmatrix} x^2 \\ y^2 \\ z^2 \end{pmatrix},$$

$$\begin{pmatrix} xy \\ -yz \\ -xz \end{pmatrix} = \begin{pmatrix} 1 & 0 & 0 \\ 0 & -1 & 0 \\ 0 & 0 & -1 \end{pmatrix} \cdot \begin{pmatrix} xy \\ yz \\ xz \end{pmatrix} \qquad (5\text{-}106)$$

$$\sigma'_{(xz)}: \begin{pmatrix} x^2 \\ y^2 \\ z^2 \end{pmatrix} = \begin{pmatrix} 1 & 0 & 0 \\ 0 & 1 & 0 \\ 0 & 0 & 1 \end{pmatrix} \cdot \begin{pmatrix} x^2 \\ y^2 \\ z^2 \end{pmatrix},$$

(5-107)

$$\begin{pmatrix} -xy \\ -yz \\ xz \end{pmatrix} = \begin{pmatrix} -1 & 0 & 0 \\ 0 & -1 & 0 \\ 0 & 0 & 1 \end{pmatrix} \cdot \begin{pmatrix} xy \\ yz \\ xz \end{pmatrix}$$

$$\sigma'_{(yz)}: \begin{pmatrix} x^2 \\ y^2 \\ z^2 \end{pmatrix} = \begin{pmatrix} 1 & 0 & 0 \\ 0 & 1 & 0 \\ 0 & 0 & 1 \end{pmatrix} \cdot \begin{pmatrix} x^2 \\ y^2 \\ z^2 \end{pmatrix},$$

(5-108)

$$\begin{pmatrix} -xy \\ yz \\ -xz \end{pmatrix} = \begin{pmatrix} -1 & 0 & 0 \\ 0 & 1 & 0 \\ 0 & 0 & -1 \end{pmatrix} \cdot \begin{pmatrix} xy \\ yz \\ xz \end{pmatrix}$$

The complete table showing one column of symbols for the representations, a central, square array of characters, and the two right columns with the bases for representation is called the character table of the group. The symbol of the group, C_{2v} in this case, is included in the upper left corner, while the symmetry operations are written at the top, across the table.

TABLE 5.7 Character table of the C_{2v} point group.

C_{2v}	E	C_2	$\sigma_v(xz)$	$\sigma'_v(yz)$		
A_1	1	1	1	1	z	x^2, y^2, z^2
A_2	1	1	-1	-1	R_z	xy
B_1	1	-1	1	-1	x, R_y	xz
B_2	1	-1	-1	1	y, R_x	yz

One may now ask whether there are new representations on other bases. To answer this important question, let us use as a basis two vectors, r_1 and r_2, representing the two O—H bonds. They will transform according to

$$E: \begin{pmatrix} r_1 \\ r_2 \end{pmatrix} = \begin{pmatrix} 1 & 0 \\ 0 & 1 \end{pmatrix} \cdot \begin{pmatrix} r_1 \\ r_2 \end{pmatrix}, \qquad \chi = 2$$

(5-109)

$$C_2: \begin{pmatrix} r_2 \\ r_1 \end{pmatrix} = \begin{pmatrix} 0 & 1 \\ 1 & 0 \end{pmatrix} \cdot \begin{pmatrix} r_1 \\ r_2 \end{pmatrix}, \quad \chi = 0 \qquad (5\text{-}110)$$

$$\sigma': \begin{pmatrix} r_1 \\ r_2 \end{pmatrix} = \begin{pmatrix} 1 & 0 \\ 0 & 1 \end{pmatrix} \cdot \begin{pmatrix} r_1 \\ r_2 \end{pmatrix}, \quad \chi = 2 \qquad (5\text{-}111)$$

$$\sigma: \begin{pmatrix} r_2 \\ r_1 \end{pmatrix} = \begin{pmatrix} 0 & 1 \\ 1 & 0 \end{pmatrix} \cdot \begin{pmatrix} r_1 \\ r_2 \end{pmatrix}, \quad \chi = 0 \qquad (5\text{-}112)$$

Clearly, the characters have the meaning of *the number of vectors left unchanged by the operations of symmetry*, and since they are invariant under any change of coordinates, we may write the apparently new representation

	E	C_2	σ	σ'	
Γ	2	0	0	2	(r_1, r_2)

This representation, however, may be reduced, by inspection of the C_{2v} character table, to the sum of two representations, namely

$$\Gamma = A_1 + B_2 \qquad (5\text{-}113)$$

It is possible to construct an infinite number of representations, which will be linear combinations of the irreducible representations (which latter are linearly independent) of the C_{2v} character table. Let us return to the equilateral triangle and construct its character table.

If we use a system of coordinates (x, y) coplanar to the triangle, the characters are those of Eqs. (5-94). The reader may verify, as a useful exercise, that the same representation is obtained on the basis of the rotations R_x and R_y. Therefore, we may write

E	$2C_3$	3σ	
2	-1	0	$(x, y), (R_x, R_y)$

The expressions in parentheses, (x, y) and (R_x, R_y), indicate that x and y (and R_x and R_y) are mixed by some of the operations of the group. If, on the other hand, we consider z and R_z, we find

E	$2C_3$	3σ	
1	1	1	z
1	1	-1	R_z

Incorporation of the quadratic forms of the coordinates finally leads to the character table of the C_{3v} group (Table 5.8).

TABLE 5.8   Character table of the C_{3v} point group.

C_{3v}	E	$2C_3$	$3\sigma_v$		
A_1	1	1	1	z	$x^2 + y^2, z^2$
A_2	1	1	-1	R_z	
E	2	-1	0	$(x, y)(R_x, R_y)$	$(x^2 - y^2, xy)(xz, yz)$

These are the three irreducible representations of the C_{3v} group. If we had used, for example, the characters of the matrices of Eqs. (5-90) we would have obtained

	E	$2C_3$	3σ
Γ	3	0	1

which is

$$\Gamma = A_1 + E. \tag{5-114}$$

If we examine the matrices of Eqs. (5-90) we see that all of them are blocked out so that they may be reduced to a number (which is in this case always $+1$ and accounts for the A_1 representation) and the group of 2×2 matrices shown in Eqs. (5-93), whose characters form the E representation.

In general, given any group of $n \times n$ matrices, they form an irreducible representation of the group if the matrices cannot be blocked out in the form of matrices of lower order ($m \times m$ matrices, with $m < n$). Conversely, if there is a change of basis (change of reference frame) for which all the matrices $\{a_{ij}\}$ may be blocked out, say, as in

$$\left(\begin{array}{c|c} \{b_{ij}\} & \{0\} \\ \hline \{0\} & \{c_{ij}\} \end{array} \right) = \mathbf{Q} \cdot \{a_{ij}\} \cdot \mathbf{Q}^{-1} \tag{5-115}$$

then the $\{a_{ij}\}$ representation is reducible.

Notice in this respect that all the matrices of the water-molecule symmetry group of Eqs. (5-97) to (5-108) are diagonal ($a_{ij} = \pm \delta_{ij}$) and therefore blocked out as E, i.e.,

$$\mathbf{E} = \left(\begin{array}{c|c|c} 1 & 0 & 0 \\ \hline 0 & 1 & 0 \\ \hline 0 & 0 & 1 \end{array} \right). \tag{5-116}$$

Consequently, all the characters of the representation are $\chi = \pm 1$, and the representations are called one-dimensional. The geometrical meaning of a symmetry group with all one-dimensional representations is that it is possible to choose a reference frame in which no operation of the group exchanges or mixes coordinates. The existence of n-fold rotations with $n > 2$ leads to coordinate exchange or mixing. In these groups, there are two-, three-, four-, and five-dimensional representations that are irreducible. It is then clear that the symmetry group of the equilateral triangle will have at least one irreducible representation that is not one-dimensional. This representation is two-dimensional, and labeled E in the character table.

In the foregoing, we have introduced two concepts: exchange and mixing of coordinates. Their meaning is different in that exchange of coordinates implies $(x \leftrightarrow \pm y)$ and any combination among x, y, z, while mixing implies $(x \leftrightarrow ax + by)$ and any combination that correlates a coordinate with a linear combination of coordinates. In rotation groups, an n-fold rotation with $n = 4$ simply exchanges coordinates, while mixing occurs for all other values of $n > 2$ if the rotation axis is parallel to one of the coordinate axes and intersects the origin.

We will now state some theorems of the theory of representations, without proof, that help in constructing the character tables.

I. **The sum of the squares of the dimensions l_i of the irreducible representations is equal to the order of the group, h, which is the number of operations of symmetry,** i.e.,

$$\sum_i l_i^2 = h. \qquad (5\text{-}117)$$

II. **The sum of the squares of the characters in any irreducible representation is equal to h,** i.e.,

$$\sum_i n_i \chi_i^2 = h \qquad (5\text{-}118)$$

where n_i is the number of operations in the class.

III. **The vectors whose components are the characters of two irreducible representations are orthogonal,** i.e.,

$$\sum_j n_j \chi_i(\mathbf{R}_j) x_k(\mathbf{R}_j) = 0. \qquad (5\text{-}119)$$

where the subscript j stands for the operation.

It should be remembered in this respect that an ordered set of h real or complex numbers to which are applied certain well-defined rules is, by definition, a vector in h dimensions. When p operators form a class, the character of the class must be written p times, so that the vector has h components, of which some of them may be zero.

IV. The number of irreducible representations of a group is equal to the number of classes.

This is equivalent to saying that it is possible to build a number of linearly independent vectors (since the representations are irreducible) equal to the number of classes. Any reducible representation will then be a linear combination of the vectors whose components are the characters of the irreducible representations.

Using these four theorems, it is possible to construct the character tables in a straightforward manner. Let us consider the symmetry group of the water molecule. We must have four irreducible representations (Theorem IV). Each one must have four characters, since the number of classes (in this case each class has one operation) is four. Since it is required by Theorem I that the sum of the squares of the dimensions of the representations be equal to the number of operations, i.e., $h = 4$, we have

$$l_1^2 + l_2^2 + l_3^2 + l_4^2 = 4. \tag{5-120}$$

Since the dimension of a representation is always a positive integer, the unique solution of Eq. (107) is

$$l_1 = l_2 = l_3 = l_4 = 1. \tag{5-121}$$

The C_{2v} group has then four one-dimensional representations. The first representation is then

	E	C_2	σ	σ'
Γ_1	1	1	1	1

for there is, in all symmetry groups, at least one basis that remains unchanged by the operations of the group. This representation satisfies Theorem II, i.e.,

$$(1)^2 + (1)^2 + (1)^2 + (1)^2 = 4. \tag{5-122}$$

Since the second representation must be one-dimensional and orthogonal to Γ_1 we readily find

	E	C_2	σ	σ'
Γ_2	1	1	-1	-1

That Theorem III is satisfied is shown by

$$(1)(1) + (1)(1) + (1)(-1) + (1)(-1) = 0. \tag{5-123}$$

Proceeding in the same manner, we finally arrive at the four irreducible representations

	E	C_2	σ	σ'
Γ_1	1	1	1	1
Γ_2	1	1	-1	-1
Γ_3	1	-1	1	-1
Γ_4	1	-1	-1	1

The identity characters obviously cannot be negative.

Comparison with the C_{2v} character table by inspection of Table 5.6 leads to

$$\Gamma_1 = A_1, \quad \Gamma_2 = A_2, \quad \Gamma_3 = B_1, \quad \Gamma_4 = B_2. \quad (5\text{-}124)$$

Finding the bases of each representation is in this case simple, as follows:

z, x^2, y^2 and z^2 remain unchanged. They form the $\Gamma_1 = A_1$ representation.

R_z and xy are changed by σ and σ' only. They form the $\Gamma_2 = A_2$ representation.

x, R_y and xz are changed by σ' and C_2 only. They form the $\Gamma_3 = B_1$ representation.

y, R_x and yz are changed by σ and C_2 only. They form the $\Gamma_4 = B_2$ representation.

Application of this procedure to the symmetry group of the equilateral triangle follows.

There are $h = 6$ operations forming three classes: E, $2C_3$, 3σ. There must be three irreducible representations. Their dimensions l_i must satisfy

$$l_1^2 + l_2^2 + l_3^2 = 6. \quad (5\text{-}125)$$

For each representation,

$$\chi^2(\mathbf{E}) + 2\chi^2(\mathbf{C}_3) + 3\chi^2(\sigma) = 6. \quad (5\text{-}126)$$

For each two representations we must have

$$\chi_i(\mathbf{E})\chi_j(\mathbf{E}) + 2\chi_i(\mathbf{C}_3)\chi_j(\mathbf{C}_3) + 3\chi_i(\sigma)\chi_j(\sigma) = 0. \quad (5\text{-}127)$$

The first representation is

	E	$2C_3$	3σ
Γ_1	1	1	1

The second is

	E	$2C_3$	$3\frac14$
Γ_2	1	1	-1

The third one must have a higher dimension, since no other combination of $\chi = \pm 1$ satisfies Theorem III. Since the dimensions must be $+1$ or $+n$ with n integer, and we already have $l_1 = 1$, $l_2 = 1$, we find $l_3 = 2$ and the third, and last, irreducible representation is

	E	$2C_3$	3σ
Γ_3	2	-1	0

Comparison with the C_{3v} character table by inspection of Table 8 leads to

$$\Gamma_1 = A_1, \qquad \Gamma_2 = A_2, \qquad \Gamma_3 = E. \qquad (5\text{-}128)$$

Finding the bases for each representation is straightforward.

A final remark about bases seems in order. The choice of x, y, z, x^2, y^2, z^2, xy, yz, and zx as bases for the representation of symmetry groups has an important bearing in the study of electronic configurations, for the p and d angular functions may be expressed as functions of the coordinates:

$$p(\pm 1) \propto x, \qquad p(\pm 1) \propto y, \qquad p(0) \propto z$$
$$d(\pm 2) \propto x^2 - y^2, \qquad d(\pm 2) \propto xy$$
$$d(\pm 1) \propto yz, \qquad d(\pm 1) \propto xz, \qquad d(0) \propto 2z^2 - x^2 - y^2$$

and transform as these latter.

Similarly, in molecular spectroscopy, it is customary to use the following bases:

a) three translation vectors T_x, T_y, T_z, that transform as x, y, z, and

b) six components of the polarizability tensor $= \alpha_{xx}$, α_{yy}, α_{zz}, $\alpha_{xy} = \alpha_{yx}$, $\alpha_{yz} = \alpha_{zy}$, and $\alpha_{zx} = \alpha_{xz}$, that transform as their subscripts,

c) the already familiar rotations R_x, R_y, and R_z.

In groups of n-fold rotations with $n > 2$ where no planes of symmetry transform C_n^m into C_n^{n-m}, C_n^m and its reciprocal C_n^{n-m} do not form a class. In such a case, it is necessary to introduce complex characters to account for the sign of each rotation. Since these representations have the same basis functions, they appear in pairs, for example in Table 5.9.

TABLE 5.9 Character table of the C_3 group.

C_3	E	C_3	C_3^2		
A_1	1	1	1	z, R_z	$x^2 + y^2, z^2$
E	$\begin{cases} 1 \\ 1 \end{cases}$	$\begin{cases} \epsilon \\ \epsilon^* \end{cases}$	$\begin{cases} \epsilon^* \\ \epsilon \end{cases}$	$\begin{cases} (x, y) \\ (R_x, R_y) \end{cases}$	$\begin{cases} (x^2 - y^2, xy) \\ (yz, xz) \end{cases}$

with $\epsilon = \exp(2\pi i/3)$ and $\epsilon^* = \exp(-2\pi i/3)$.

Since

$$\epsilon = \exp(i\theta) = \cos \theta + i\sin \theta \qquad (5\text{-}129)$$

$$\epsilon^* = \exp(-i\theta) = \cos \theta - i\sin \theta \qquad (5\text{-}130)$$

which are the familiar forms of two complex numbers that are conjugate to each other, these exponential functions reduce to $\pm i$ for $n = 4$, so that the character table of the C_4 group (Table 5.10) has characters that are ± 1 or $\pm i$.

TABLE 5.10   Character table of the C₄ group.

C_4	E	C_4	C_2	C_4^3		
A	1	1	1	1	z, R_z	$x^2 + y^2, z^2$
B	1	-1	1	-1		$x^2 - y^2, xy$
E	$\begin{cases} 1 \\ 1 \end{cases}$	$\begin{matrix} i \\ -i \end{matrix}$	$\begin{matrix} -1 \\ -1 \end{matrix}$	$\begin{matrix} -i \\ i \end{matrix}$	$\begin{cases} (x, y) \\ (R_x, R_Y) \end{cases}$	(yz, xz)

In imaginary characters, i stands for a counterclockwise rotation by $\tilde{\pi}/2$ radians and $-i$ for the clockwise rotation, while $\chi(C_2) = -1$ stands for rotation by π radians in the E representation, since $i^2 = (-i)^2 = -1$, C_2 being $C_2 = C_4^2 = (C_4^3)^2$.

In physical problems, however, it is convenient to combine the two parts of the E representation. In the case of the C_3 groups, for example, this procedure leads to

$$E \begin{array}{c|ccc} & 1 & \epsilon & \epsilon^* \\ & 1 & \epsilon^* & \epsilon \\ \hline E & 2 & 2\cos(2\pi/3) & 2\cos(2\pi/3 \end{array}$$

since the sum of two complex numbers that are conjugate to each other is a real number [Eqs. (5-129) and (5-130)] and $\cos(4\pi/3) = \cos(-2\pi/3) = \cos(2\pi/3)$. Clearly, we find in the combined E representation the character of the three-fold rotation matrix derived in Eqs. (5-93), since $\cos(2\pi/3) = -\frac{1}{2}$. The simplified character table of the C_3 group then becomes

	E	C_3	C_3^2
A	1	1	1
E	2	-1	-1

with the bases displayed in Table 5.8. Notice, however, that this way of presenting the E representation is mathematically incorrect (it does not satisfy Theorem I) and only justified for its physical applications.

5.16 Mulliken symbols

So far, we have labeled irreducible representations with capital letters and subscripts. These symbols were introduced by Mulliken with the meaning that follows.

Letters A and B stand for all one-dimensional representations. Representations which are symmetric with respect to rotations about the principal axis $[\chi(\mathbf{C}_n) = +1]$ are designated A, otherwise $[\chi(\mathbf{C}_n) = -1]$ they are designated B. Subscripts $_1$ and $_2$, attached to A and B, designate representations that are symmetric ($\chi = 1$) or antisymmetric ($\chi = -1$) with respect to rotations about a two-fold axis perpendicular to the principal n-fold axis, or to a symmetry plane containing the $\mathbf{C}_n$-axis (σ_v or σ_d) if the $\mathbf{C}_2$-axes are lacking. Two- and three-dimensional representations are respectively designated E and T (sometimes F). Symbols G and H, found in the icosahedral group, designate four- and five-dimensional representations, respectively. Subscripts g (German *gerade*, even) and u (German *ungerade*, odd) are used in groups with center of symmetry to designate representations with $\chi(i) = 1$ and $\chi(i) = -1$, respectively. Primes and double primes are attached to all letters, when applicable to indicate symmetry and antisymmetry with respect to a plane perpendicular to the principal rotation axis (σ_h). Numerical subscripts attached to E and T will not be considered, for such a discussion would be beyond the scope of this book.

General References

Cotton, F. A., "Chemical Applications of Group Theory," New York, Interscience Publishers, 1963.*

Eyring, W., Walter, J., and Kimball, G. E., "Quantum Chemistry," New York, John Wiley & Sons, 1944.

Jaswon, M. A., "An Introduction to Mathematical Crystallography," London, Longmans, 1965.

Margenau, H., and Murphy, G. M., "The Mathematics of Physics and Chemistry," 2nd. ed., Princeton, Van Nostrand, 1956.

Schonland, D. S., "Molecular Symmetry," New York, Van Nostrand, 1965.*

Wilson, E. B., Jr., Decius, J. C., and Cross, P. C., "Molecular Vibrations," New York, McGraw-Hill, 1955.

Zachariasen, W. H., "Theory of X-ray Diffraction in Crystals," New York, John Wiley & Sons, 1945.*

*Especially recommended.

Measurement of magnetic susceptibilities

6 *Experimental methods*

6.1 The Gouy method

The majority of magnetic susceptibility measurements is performed by determining the apparent change in the weight of a sample due to application of a magnetic field. Some electromagnetic devices have also been developed in which is measured the change of inductance of a coil upon introduction of the sample. This section will develop the principles of operation of general magnetogravimetric methods, particularly the Gouy method.[12]

The difference in magnetic potential energy per unit volume between a sample of permeability μ and the displaced medium of permeability μ_o is

$$- (\mu - \mu_o) H^2/8\pi \qquad (6\text{-}1)$$

The physical meaning of the negative sign is that one should either have to do work to turn the spins with respect to the lines of force of the magnetic field or to supply thermal energy in order to decrease the magnetization of the material by increasing the thermal agitation. Since the permeability μ is related to the volume susceptibility κ by

$$\mu = 1 + 4\pi\kappa \qquad (6\text{-}2)$$

Eq. (6-1) may also be written, in terms of volume susceptibilities, as

$$-\frac{1}{2}(\kappa - \kappa_0)H^2 \qquad (6\text{-}3)$$

The susceptibility can be, and usually is, referred to unit mass instead of unit volume. It is then symbolized χ and related to κ by

$$\kappa = \rho\chi \qquad (6\text{-}4)$$

where ρ is the density of the material. Though expression of results in units of χ is found more often in the literature, the quantity that has physical meaning is κ, for the natural expression of the magnetic energy is per unit volume. The mass susceptibility χ gains physical meaning when referred to one gram atom, gram ion, or gram molecule, as already seen in the theoretical treatment of paramagnetism and derivation of Curie's law.

In view of Eq. (6-3), the force exerted in the direction s upon a volume element dV is

$$dF_s = -\frac{d}{ds}\left[-\frac{1}{2}(\kappa - \kappa_0)H^2\right]dV$$

$$= \frac{1}{2}(\kappa - \kappa_0)\frac{dH^2}{ds}dV = (\kappa - \kappa_0)H\frac{dH}{ds}dV \qquad (6\text{-}5)$$

Since

$$H^2 = H_x^2 + H_y^2 + H_z^2 \qquad (6\text{-}6)$$

Eq. (6-5) can be written as

$$dF_s = (\kappa - \kappa_0)\left(H_x\frac{dH_x}{ds} + H_y\frac{dH_y}{ds} + H_z\frac{dH_z}{ds}\right)dV \qquad (6\text{-}7)$$

Eq. (6-7) may be simplified if the experimental conditions are set in order to have, for example,

$$s = z, \qquad H_y = H_z = 0 \qquad (6\text{-}8)$$

for in such a case it reduces to

$$dF_s = (\kappa - \kappa_0)H\frac{dH}{dz}dV \qquad (6\text{-}9)$$

Fulfillment of conditions (6-8) is readily achieved if the sample is held free to displace vertically (z-axis) and placed in the median plane of the pole pieces with their axis of symmetry horizontal, for in the median plane the magnetic lines of force are then perpendicular to the z-axis. We arbitrarily call the symmetry axis x-axis, and have $H_y = H_z = 0$.

In order to integrate Eq. (6-9) and obtain the total force acted on the sample, one must impose certain conditions on the specimen shape. If the specimen is cylindrical, of cross section A and length L, and is suspended with its long axis vertical, Eq. (6-9) can be readily integrated to give

$$F = (\kappa - \kappa_0) A \int_{H_0}^{H_L} H \, dH = \frac{1}{2} (\kappa - \kappa_0) A (H_L^2 - H_0^2) \qquad (6\text{-}10)$$

where F and z have the same sign when $\kappa - \kappa_0 > 0$, if the sense of the z-axis is taken positive *downwards* and F is measured by weighing. If the cross section of the sample, A, is small (as compared to the gap between the pole pieces) and the sample is correctly centered in the median plane, it can be assumed that

$$\frac{dH_x}{dz} = \frac{dH_y}{dz} = 0 \qquad (6\text{-}11)$$

We will return to Eq. (6-11) later in discussing the causes of error in the determination of susceptibilities using this method. Fig. 6.1 shows a schematic diagram of such an arrangement. If the sample is now weighed in the presence and in the absence of magnetic field, and $g\delta m$ is the apparent difference in weight, then

$$g\delta m = \frac{1}{2} (\kappa - \kappa_0) A (H_L^2 - H_0^2) \qquad (6\text{-}12)$$

This is the principle of operation of the method known under the name of Gouy, who introduced it at the end of the nineteenth century.[12]

Two advantageous features of this method are that if the diameter of the pole pieces is large with respect to their separation, the position of the bottom of the sample is not critical because the field is, for all practical purposes, uniform in a large region around the center; if the sample is long as compared to the diameter of the pole pieces, the magnetic field at the top may be neglected with respect to the magnetic field at the bottom. Eq. (6-12) then reduces to

$$g\delta m = \frac{1}{2} (\kappa - \kappa_0) A H_L^2 \qquad (6\text{-}13)$$

The main disadvantage of the Gouy method is the rather large amount of sample needed.

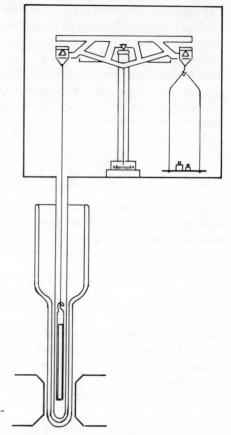

FIG. 6-1 Gouy device for the determination of magnetic susceptibilities.

In order to proceed further, we will make another simplifying assumption: that the susceptibility of the displaced medium can be neglected when compared with the susceptibility of the sample, i.e.,

$$\kappa_o \ll \kappa \tag{6-14}$$

and write Eq. (6-13) in the simplest form

$$g\delta m = \frac{1}{2}\kappa A H_L^2 \tag{6-15}$$

Such an assumption is not very restrictive when measuring solid or liquid samples. Even if the displaced medium were air—which is paramagnetic due to its oxygen content—the neglected value would be 0.03×10^{-6} emu at NTP, which, when compared with volume susceptibility values of various materials in condensed phases, is indeed small. Let alone that one could always use a nitrogen atmosphere, whose susceptibility is -6×10^{-9} emu at NTP, helium, or even a vacuum.

Eq. (6-15) may be applied to liquid samples and to cylinder-shaped solids. In dealing with powdered solids, the value of ρ looses its meaning since it must be referred to the volume of the sample, which clearly depends on the grain size and on the degree of compactness. If one assumes uniform compactness, one may establish the correspondence

$$dV = A\,dz = \rho^{-1}dm \qquad (6\text{-}16)$$

as constant along the length L of the cylinder. The value of ρ is, in this case, the apparent density of the material, equal to

$$\rho = \frac{m}{AL} \qquad (6\text{-}17)$$

where m is the mass of the sample and AL its volume. Elimination of A between Eqs. (6-15) and (6-17) and consideration of Eq. (6-4) finally leads to

$$g\delta m = \tfrac{1}{2}\kappa A H_L^2 = \tfrac{1}{2}(\kappa m/\rho L)H_L^2 = \tfrac{1}{2}\chi(m/L)H_L^2 \qquad (6\text{-}18)$$

which permits the evaluation of the mass susceptibility from

$$\chi = \frac{2Lg}{H_L^2}\frac{\delta m}{m} \qquad (6\text{-}19)$$

Eq. (6-19) may be used to evaluate χ if the value of H_L is determined by independent means. Otherwise, the device must be calibrated with a substance of known magnetic susceptibility χ_o, in which case one readily arrives, after eliminating H_L, g, and L, at

$$\chi = \frac{m_o\,\delta m}{m\,\delta m_o}\chi_o \qquad (6\text{-}20)$$

where the subscript o labels the variables that belong to the reference substance. Standard substances used for calibration are: oxygen in the case of gases, water and solutions of nickel chloride in the case of liquids, and powdered crystals of $Mn\,SO_4 \cdot 4H_2O$ or $Fe(NH_4)_2(SO_4)_2 \cdot 6H_2O$ and, in the case of solids, a few other well known crystals.

The results from Eqs. (6-19) and (6-20) *do not depend on the degree of compactness*, provided this be uniform along the tube. It is now advisable to recapitulate and see where assumptions have been made and to what extent they could introduce errors in measurement. The assumptions made are:

(1) uniform field at the bottom of the sample
(2) constant cross section A

(3) $dH_x/dz = dH_y/dz = 0$ (Eq. 6-11)
(4) $H_L \gg H_o$ (Eq. 6-13)
(5) $\kappa \gg \kappa_o$ (Eq. 6-14)
(6)   uniform compactness
(7)   implicit assumptions in the derivation of Eq. (6-20).

Assumption (1) is easily satisfied with suitable electromagnets. Assumption (2) can always be justified if a precision-bored tube is used, whose bottom end can be cut in a lathe and covered with a flat plate, appropriately sealed. Assumption (4) is secured by using long samples. Assumption (5) has already been discussed, and the correction for the susceptibility of the medium can always be made, if needed. Assumptions (3), (6), and (7) need more careful consideration.

With respect to Assumption (3), the terms dH_x/dz and dH_y/dz cannot always be ignored. Careful as the arrangement may be, small values are always present to some extent. Although they need not be considered for the purpose of calculating the susceptibility, they may cause disturbances that add up to the difficulties of the method. These disturbances take the form of lateral displacements of the sample, which are enormously increased by the presence of ferromagnetic impurities in the specimen. Such displacements may become particularly cumbersome when the sample is mounted inside a narrow tube, as is usually the case. The tube may be the wall of a thermostat or of a draught-proof case. It is clear that any lateral displacement that may bring the sample into contact with the wall will lead to irreproducible results. Such displacements are greatly reduced if the sample is suspended with its upper end in the strong field and its lower end outside the field, below the pole pieces.

The problem of compactness—Assumption (6)—is still open to discussion. In the author's experience, better than one in one hundred accuracies may be reached if the simple but effective technique that follows is used. The crystals are finely ground in an agathe mortar, and the sample tube filled by adding a small amount of material at a time, which is then thoroughly compacted by repeatedly dropping the tube from a certain height, inside a larger and longer tube, until the desired level is reached. The reader will recognize in this technique the one used to fill capillaries for measuring melting point of organic compounds.

We now arrive at (7), regarding implicit assumptions made in the derivation of Eq. (6-20). In order to eliminate L, g, and H_L by measuring a standard susceptibility, their values must be accurately reproduced in both cases, i.e., standard and unknown. Since the gravity acceleration g is a constant in each place, we need not be concerned about it. The variables L and H_L are, on the contrary, likely to change; the first due to failure in filling both tubes to the same level, the second due to lacking of appropriate field stabilization. While the error introduced by L only appears in powdered samples for which Eq. (6-20) must be used, the error arising from an unstable field is general and should be carefully considered. Commercial equipments may be provided with highly stabilized fields but this option is, of course, dearly bought. Finally, the total magnetic pull $g\delta m$ includes the effect of the magnetic field on the container and should be corrected for it. This can be done by making a separate experiment with the empty reservoir. An alternative means is to use a tube divided into two equal compartments, the upper containing the sample. The division wall should be placed in the middle of the pole pieces. Due to symmetry considerations, the magnetic pull on the container cancels out and $g\delta m$ then represents the magnetic force exerted on the sample alone.

The Gouy method has extensively been used and modified to fulfill specific purposes. Among the modifications, perhaps the most important is the one known as Quincke method;[22] an excellent example of application is the work of Auer.[1] This method is applicable to liquids only. The liquid under study is placed inside a U-shaped glass reservoir. One arm is a narrow, precision-bored tube, while the other is very wide. The narrow limb is placed in the median plane of the magnet pole pieces so that the surface of the liquid lies close to the center in the absence of a magnetic field. The wide limb is outside the magnet, in a region where the magnetic field may be neglected. Upon excitation of the electromagnet, the magnetic pull on the liquid displaces the level in the narrow limb without appreciable change in the wide one until equilibrium with the hydrostatic pressure is reached; i.e.,

$$\frac{1}{2}(\kappa - \kappa_0)AH^2 = A(\rho - \rho_0)gh \qquad (6\text{-}21)$$

where ρ is the density of the liquid; the subscript o now stands for the displaced medium; and h is the difference in level between both arms.

Although magnetic susceptibility measurements are rather difficult, it is possible to make fairly good determinations in an introductory course with a simplified device using a Westphall balance.[17] Fig. 6.2 illustrates the experimental setup. Since both the mass m and the apparent change δm due to the force exerted by the magnetic field may be measured in the conventional scale of the riders—as long as Eqs. (6-19) and (6-20) only include mass ratios—it is not necessary to know the mass of the unit rider. With fields slightly above 6000 Oersted, readily attained with a small electromagnet, one can determine susceptibilities with an absolute error of the order of 2×10^{-7} emu/g. Although this error is large with respect to diamagnetic susceptibilities, it permits, at room temperature, the de-

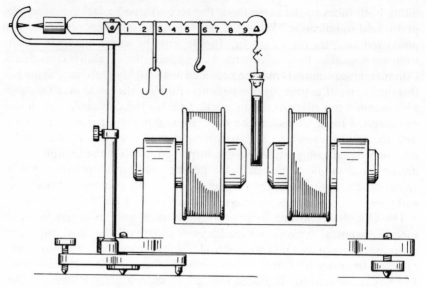

FIG. 6-2 Gouy device for class demonstration.

termination of susceptibilities of strongly paramagnetic samples with a relative error of a few percent.

6.2 The Faraday method

In considering the magnetic pull on the volume element dV given by Eq. (6-9)

$$dF_z = (\kappa - \kappa_0) H \frac{dH}{dz} dV \qquad (6\text{-}9)$$

the problem can be simplified in a different way than that used in Section 6.1—namely, by choosing experimental conditions so that the expression $H(dH/dz)$ be a constant over the volume V of the sample, for, in such a case, integration of Eq. (6-9) is straightforward, giving

$$g\delta m = (\kappa - \kappa_0) H \frac{dH}{dz} V \qquad (6\text{-}22)$$

where the magnetic pull is again weighed. Even if the sample is small, and hence its volume V, the method requires the value of $H(dH/dz)$ to be reasonably constant over a certain distance along z, as well as dH_x/dz, and dH_y/dz as close to zero as possible. This method was developed by

Faraday[5] in his epoch-making research on the magnetic properties of matter. Pole tips of special shapes that ensure the constancy of $H(dH/dz)$ over one or two centimeters have been designed[7] and are available in the market. Fig. 6.3 shows a schematic diagram of a device of this type. Clearly, the advantage of this method over Gouy's lies in the small amount of sample needed for a determination: usually not more than 100 mg, and often less. An interesting feature of this method is that the magnetic pull only depends on the mass of material. In fact, if one combines Eqs. (6-22) and (6-4), κ and V are eliminated, arriving at

$$g\delta m = (\chi - \chi_0) H \frac{dH}{dz} m \qquad (6\text{-}23)$$

which does not depend on the density provided the volume of the sample be confined in the region of constant $H(dH/dz)$. Since the sample is so small, much variety is found in the devices used to measure the magnetic pull, although the most commonly used is a conventional microbalance. Commercial equipment of this type is available in the market. The ulti-

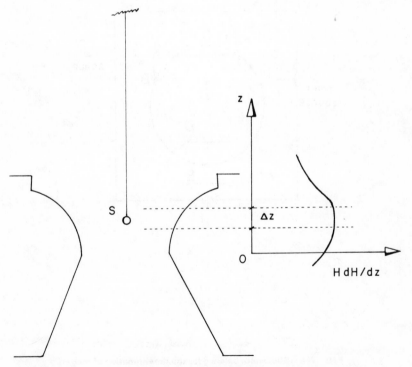

FIG. 6-3 Faraday device for the determination of magnetic susceptibilities.

mate sensitivity exceeds that of any other type of susceptibility measurements (with the exception of the less quantitative electron paramagnetic resonance). With a microbalance having a sensitivity of 0.1 microgram and commercially available electromagnets, a mass susceptibility change of 10^{-10} emu/g may be measured. If the device is very sensitive, it is advisable to use a beta emitter for eliminating static charges that may develop in the sample and the inner wall of the jacket.

In addition to the conventional microbalance and the torsion balance, two interesting modifications merit attention: the Sucksmith ring balance and the Föex and Forrer translation balance.

The Sucksmith balance,[27] diagramatically shown in Fig. 6.4, consists of a ring made of phosphor bronze strip, rigidly fixed at O with its plane vertical. Two mirrors, C and B, and a pan P attached to the sample suspension F are also rigidly fixed to the ring. The device is inside a box, and a vane E provides the necessary dampening of the oscillations started by excitation of the electromagnet. A beam of light enters through lens L, is reflected by the mirrors, and leaves the box to form an image on a scale.

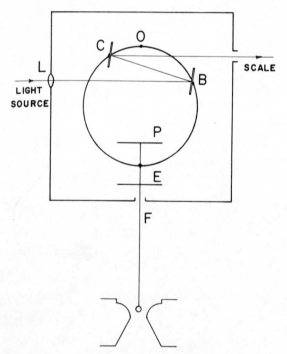

FIG. 6-4 Sucksmith balance for the determination of magnetic susceptibilities.

The deformation of the ring due to the magnetic pull is greatly amplified by the double reflection and permits a rapid measurement of susceptibilities that are not small. The pan P allows calibration of the scale in mass units, and Eq. (6-23) is used.

In the Föex and Forrer balance,[8] schematically shown in Fig. 6.5, the sample S is attached at one end of a light diamagnetic metal rod, while the other end of the rod has a coil C lying in the gap of a permanent magnet,

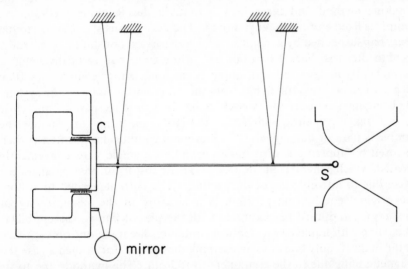

F I G. 6.5 Föex and Forrer balance for the determination of magnetic susceptibilities.

like the coil of a loudspeaker. The rod is suspended in such a way that it can only move along its axis, centered in the median plane of the irregularly shaped pole tips of an electromagnet. The sample is located inside the region of constant $H(dH/dz)$. When the electromagnet is excited, the magnetic pull exerted on the sample pulls the rod inside the pole gap as the coil at the other end separates from the permanent magnet, or conversely. A variable voltage is then applied to the coil until the original equilibrium position is restored. There is a linear correspondence between the current I circulating through the coil and the magnetic pull on the sample, i.e.,

$$KI = (\chi - \chi_0)H \frac{dH}{dz} m \qquad (6\text{-}24)$$

where K is a constant characterizing the device.

6.3   The Rankine method

Diamagnetic and weakly paramagnetic samples present difficult experimental problems when Gouy and Faraday methods are used. In the particular case of liquids, however, we have seen that Quincke's adaptation of Gouy's method may be accurate. A simpler, though extraordinarily sensitive method was developed for this purpose by Rankine.[23] The theory of the method is very involved; the apparatus must be calibrated with a liquid of known susceptibility. The main difference between the Rankine method and the others is that, in the Rankine method, the sample is fixed and the magnet moves. The device consists of a horizontal cross bar suspended by a metal strip of low torsion constant. A mirror is fixed to the cross bar. From one end of the cross bar vertically hangs a permanently magnetized rod, about 5 cm long, and weighing less than 10 g. An equal brass rod hangs from the other end, balancing the system. On bringing a cylinder of a weakly magnetic substance parallel and close to the magnet, it attracts or repels the latter according to whether the sample is para or diamagnetic. The original equilibrium position is then restored by applying a very small current to a single wire conveniently located, which generates the desired magnetic distortion that balances the effect due to introduction of the sample. The current needed to restore the original equilibrium position is a measure of the susceptibility of the sample, and must be calibrated with samples of known susceptibility. In dealing with liquids, two identical containers are used, one on each side of the magnet, only one of them carrying the sample, for in such a case the magnetic pulls due to the containers cancel out. The method, due to its very high sensitivity, and the moving magnet, is particularly sensitive to spurious magnetic fields, including the earth vertical magnetic component. What is then gained by the simplicity of the device is lost by the necessity of providing the installation with a special magnetic environment. The method, however, has been successfully used for comparing the magnetic behavior of light and heavy water.[16]

6.4   Correction for ferromagnetic impurities

The presence of ferromagnetic impurities may not only cause the sample to be more sensitive to the field asymmetries discussed in Section 6.1, but may also distort the results. Ferromagnetic impurities may appear in the sample as well as in the container, in this latter mainly as contamination with atmospheric dust. It is difficult to determine the precise way in which ferromagnetic impurities in the bulk of the sample may affect the results, for the ferromagnetic character is intimately related with the size of the

impurity if it is finely divided and spread throughout the body of the specimen. Also, if the impurity is dissolved but tends to segregate in clusters,
the situation may be very different of that presented by a homogeneous
solid solution. Small amounts of iron in copper and larger amounts in
aluminum have been claimed not to behave ferromagnetically. The
presence of ferromagnetism, on the other hand, can be detected in frozen
iron amalgams having as little as one part per million of iron. Honda[15]
and Owen[20] have studied the problem if the contamination is not large.
The basic assumption behind their treatment is that saturation of the ferromagnetic impurity occurs, in which case the force exerted on the impurity must be proportional to dH/dz, while the force acting on para and
diamagnetic bodies is, in general, proportional to $H(dH/dz)$. The ob-
served susceptibility κ_H is then connected to the true susceptibility κ_∞ by

$$\kappa_\infty = \kappa_H - c\sigma_s/H \qquad (6\text{-}25)$$

where c is the concentration in g/cm^3 of the ferromagnetic impurity and
σ_s its specific magnetization of saturation. Vogt[28] showed that Eq. (6-25)
could be used for Faraday method, but that for Gouy's the equation
should be

$$\kappa_\infty = \kappa_H - 2c\sigma_s/H \qquad (6\text{-}26)$$

Clearly, determination of the susceptibility at various magnetic fields
and further plotting of κ_H vs. $1/H$ makes the correction trivial.

However, the derivation of Eqs. (6-25) and (6-26) is based in the rather
restrictive assumption that saturation actually occurs. This may not be
true for certain impurities, or if the value of the magnetic field is not large
enough. Bates and Baker[2] have derived a more elaborate correction for
the Gouy method, in which the demagnetization coefficient D and the
saturation intensity of magnetization I_s of the impurity are taken into
account:

$$\kappa_H = \kappa_\infty + \frac{2c\sigma_s}{H} - \frac{c\sigma_s DI_s}{H^2} \qquad (6\text{-}27)$$

Eq. (6-27) is valid provides that $(3h + DI_s) < H$, where h is the coerci-
tivity of the impurity.

6.5 Miscellaneous methods

For ordinary purposes, the Gouy and Faraday methods suffice, with all
the variations adopted, to fulfill most needs. More difficult problems re-
quire ingenious, sometimes highly sophisticated devices. It would be out

of the scope of this book to review all these methods in detail. A word should be said, however, about the determination of gaseous susceptibilities. With a few exceptions (O_2, NO, NO_2, ClO_2), gases are diamagnetic. Since the magnetic pull actually depends of the volume susceptibility, the pull exerted on gaseous samples is very weak, for gases have densities in molecules per unit volume three orders of magnitude below those of condensed phases. Extremely sensitive devices have been developed for this purpose. In some of them the pull on the gas is directly measured.[25] In others, what is measured is the change is the magnetic pull on a standard body immersed in different gases,[26] or the change in the magnetic torque that acts on bodies of heterogeneous susceptibility under similar conditions.[3,11,13].

Rather recently, a very ingenious method that makes use of the high resolution of nuclear magnetic resonance has been developed for measuring the susceptibility of solid and liquid paramagnetic substances.[6]

6.6   Treatment of data

From what has been seen in Chapter 4, data on paramagnetic susceptibilities are of little value if they are not taken at several temperatures. For this reason, various devices have been described for measuring susceptibilities from very low to very high temperatures. The problem that we must discuss now is what to do once these data have been collected. It will be assumed that the correction for ferromagnetic impurities described in Section 6.4, if necessary, has been performed. The diamagnetic contribution may be subtracted (its absolute value added) by adopting experimental values of isoelectronic compounds. The value of the obtained susceptibility can then be plotted as a function of the inverse absolute temperature. In the case of ideal paramagnetic behavior, the plot χ vs. $1/T$ must be a straight line passing through the origin, since Curie's law is obeyed, i.e.,

$$\chi = C/T \tag{6-28}$$

If the plot is a straight line, but does not go through the origin, two situations may arise: either the straight line, extrapolated at $1/T = 0$, intersects the positive χ-axis (one must suspect either contribution of temperature-independent paramagnetism, or overestimation of the diamagnetic correction) or the extrapolated value intersects the negative χ-axis (in this case, the diamagnetic correction has, most likely, been insufficient). If the plot is not a straight line, a Curie-Weiss law

$$\chi = \frac{C}{T + \Delta} + a \tag{6-29}$$

or a Néel law

$$\chi = \frac{C}{T - \Theta} + a \tag{6-30}$$

must be tried. The parameters Δ and Θ are called the Weiss and Néel constants.

The alternative plot χT *vs.* T must be interpreted as follows. A Curie law gives a straight line parallel to the T-axis. A straight, but sloping line indicates one of the above cases, according to the sign of the slope. A curved line corresponds to Curie-Weiss or Néel behavior. In the latter cases, it is possible to derive the values of all constants of Eqs. (6-29) and (6-30) by successive approximations.[19]

If Curie's law is followed, the results may be expressed in effective number of magnetons (see Section 3.4):

$$p_{eff.} = 2.839 \, C^{1/2} \tag{6-31}$$

A unit of atomic moments with no theoretical significance, the *Weiss magneton*, is still found in the literature. The effective number of Weiss magnetons is given by

$$p_{\rm W} = 14.05 \, C^{1/2} \tag{6-32}$$

The evaluation of the probable error affecting the value of $p_{eff.}$ is a trivial problem if the errors of both temperature and magnetic pull are known. Large errors and lacking of reproducibility are mainly due to absence of thermal equilibrium, sample in occasional contact with the walls, and electrostatic charges in the system. However, very accurate susceptibility measurements can be made with a good deal of experience, even with limited resources.

6.7 Electromagnetic methods

The inductance L of a coil immersed in a medium of magnetic susceptibility κ is

$$L = L_0(1 + 4\pi\kappa) \tag{6-33}$$

where L_0 is its inductance in a vacuum. If the sample is placed inside the coil, the inductance L is given, neglecting the susceptibility of the displaced medium, by

$$L = L_0(1 + 4\pi\xi\kappa) \tag{6-34}$$

where ξ is a filling factor less than unity. If the coil is made to oscillate as a part of an LC circuit, the following equation will relate the oscillation frequencies f and f_o, with and without sample

$$f(1 + 4\pi\xi\kappa)^{1/2} = f_o \tag{6-35}$$

where the influence of the displaced medium has again been neglected. Since $4\pi\xi\kappa$ is much smaller than unity (under ordinary conditions), Eq. (6-35) may be simplified to

$$f = f_o(1 - 2\pi\xi\kappa) \tag{6-36}$$

If another circuit, oscillating at f_o, is available, it is easy to measure $f_o\text{-}f$ by conventional electronic means, which leads to

$$\kappa = \frac{f_o - f}{2\pi\xi f_o} \tag{6-37}$$

The factor ξ must be determined by calibrating the device with a sample of known paramagnetism. Since κ is of the order of 10^{-6} to 10^{-4} emu/cm,[3] the frequency relative change is also 10^{-6} to 10^{-4}, because $2\pi\xi$ is of order unity.

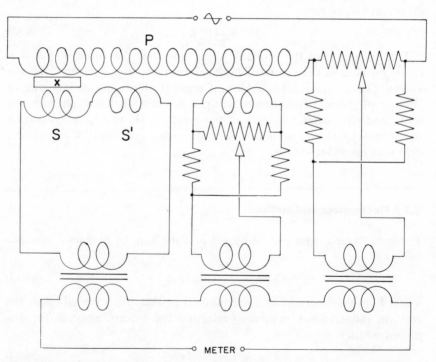

FIG. 6-6 Schematic inductance bridge for the determination of magnetic susceptibilities.

An alternate and somewhat simpler procedure is to measure the inductance L in an inductance bridge. This method is widely used in low-temperature techniques.[10]

The methods described above are, in practice, limited in accuracy to values of either $(f_0 - f)/f_0$ or $(L_0 - L)/L_0$ not much smaller in absolute value than 10^{-6}. Measurement of mutual inductance may provide greater accuracy. A schematic drawing of a setup is shown in Fig. 6.6. This apparatus essentially consists of a primary coil P producing a homogeneous, variable magnetic field in a large region. Two secondary coils, S and S', wound up in opposition and connected in series, generate an induced voltage that would be balanced if the coils were identical. Since this condition cannot be fulfilled in practice, and because there are always losses that produce a phase shift, the balance in phase and amplitude must be obtained by a derivation of the primary current and an additional secondary coil, as indicated. Introduction of the sample X in one of the secondary coils unbalances the bridge, which must be reset both in phase and amplitude. The new settings, if the device is properly calibrated, will be a measure of the susceptibility. The theory and details of construction and operation of this and other related devices is out of the scope of this book and should be consulted in the pertinent literature.[4,14,18,24] A beautiful application of this method has been made for measuring the concentration of nitrogen atoms trapped in solid molecular nitrogen at $4°K$.[9]

References

1. Auer, H., *Ann. Physik* **18**, 593 (1933).
2. Bates, L. F., and Baker, J. W., *Proc. Phys. Soc.* **52**, 443 (1940).
3. Bitter, F., *Phys. Rev.* **35**, 1572 (1930).
4. DeKlerk, D., and Hudson, R. P., *J. Res. Natl. Bur. Standards* **53**, 173 (1954).
5. Faraday, M., "Experimental Researches," Vol. 3, pp. 27 and 497, London, Taylor and Francis, 1855.
6. Feher, G., and Knight, W. D., *Rev. Sci. Instr.* **26**, 293 (1955).
7. Fereday, R. A., *Proc. Phys. Soc.* **43**, 383 (1931).
8. Föex, G., and Forrer, R., *J. Phys. Radium* **7**, 180 (1936).
9. Fontana, B. J., *J. Chem. Phys.* **31**, 148 (1959).
10. Giauque, W. F., and MacDougall, D. P., *J. Amer. Chem. Soc.* **57**, 1175 (1935).
11. Glaser, A., *Ann. Physik* **75**, 459 (1924).
12. Gouy, L. G., *C. R. Acad. Sci. Paris* **109**, 935 (1889).
13. Hammar, G. W., *Proc. Natl. Acad. Sci., Wash.,* **12**, 597 (1926).
14. Heukelom, W., Broeder, J. J., and van Reijen, J., *J. Chim. Phys.* **51**, 474 (1954).

15. Honda, K., *Ann. Physik* **32**, 1048 (1910).
16. Iskerendian, H. P., *Phys. Rev.* **51**, 1092 (1937).
17. McMillan, J. A., *Amer. J. Phys.* **27**, 352 (1959); ibid. **29**, 207 (1961).
18. Mesnage, M. C., Grivet, P., and Souzade, M., *C. R. Acad. Sci. Paris* **249**, 59 (1959).
19. Néel, L., *Ann. Phys. Paris* **5**, 232 (1936).
20. Owen, M., *Ann. Physik* **37**, 657 (1912).
21. Pacault, A., Hoarau, J., and Marchand, A., *Advan. Chem. Phys.* **3**, 171 (1961).
22. Quincke, G., *Ann. Physik* **24**, 347 (1885); ibid. **34**, 401 (1888).
23. Rankine, A. O., *Proc. Phys. Soc.* **46**, 391 (1934).
24. Selwood, P. W., "Adsorption and Collective Paramagnetism," p. 89, New York, Academic Press, 1962.
25. Sone, T., *Phil. Mag.* **39**, 305 (1930).
26. Stössel, R., *Ann. Physik* **10**, 393 (1931).
27. Sucksmith, W., *Phil. Mag.* **8**, 158 (1929).
28. Vogt, E., *Ann. Physik* **14**, 1 (1932).

General references

Bates, L. F., "Modern Magnetism," 4th ed., Cambridge, University Press, 1961.
Selwood, P. W., "Magnetochemistry," 2nd ed., New York, Interscience, 1956.

7 *Measurement on single crystals*

7.1 Anisotropic magnetization

In Chapter 6 we assumed that the samples were magnetically isotropic, i.e., that the magnetic susceptibility could be expressed by a scalar quantity. Such an assumption is justified in two cases: first, in measurements performed on powdered, polycrystalline, and non-crystalline samples, and secondly, in isotropic single crystals. Since only cubic crystals are isotropic, a discussion of measurements on single crystals must begin with a discussion of the type of behavior of crystals of other systems. Insisting on what was said in Chapters 1 and 6, magnetic anisotropy is the property of some crystals of becoming magnetized in a direction not necessarily parallel to the intensity vector of the applied magnetic field. The susceptibility takes the form of a symmetric second-rank tensor and may be expressed by a (3×3) matrix. If H_x, H_y, H_z are the components of the magnetic vector $\mathbf{H}$ in an arbitrary frame, then, in general, the components of the magnetization vector **M** will be given by

$$M_x = \kappa_{11}H_x + \kappa_{12}H_y + \kappa_{13}H_z$$
$$M_y = \kappa_{21}H_x + \kappa_{22}H_y + \kappa_{23}H_z \qquad (7\text{-}1)$$
$$M_z = \kappa_{31}H_x + \kappa_{32}H_y + \kappa_{33}H_z$$

with $\kappa_{ij} = \kappa_{ji}$. Eqs. (7-1) may be put in symbolic form as

$$\mathbf{M} = \boldsymbol{\kappa} \cdot \mathbf{H} \qquad (7\text{-}2)$$

where **H** and **M** are column vectors and $\boldsymbol{\kappa}$ is a matrix of terms

$$\boldsymbol{\kappa} = \{\kappa_{ij}\} \qquad (7\text{-}3)$$

It is generally possible to find a reference frame with respect to which the nondiagonal terms of Eq. (7-3) vanish. The three orthogonal axes of such a frame are termed the *principal susceptibility axes*. The diagonal matrix is then

$$\boldsymbol{\kappa} = \begin{bmatrix} \kappa_1 & 0 & 0 \\ 0 & \kappa_2 & 0 \\ 0 & 0 & \kappa_3 \end{bmatrix} \qquad (7\text{-}4)$$

where the nomenclature κ_i is used for the diagonal terms k_{ii} after diagonalization. The values κ_i are called *principal susceptibilities*. Since a suitable change of coordinates is necessary in order to diagonalize the susceptibility matrix, the diagonal matrix looked for is a similarity transform of the nondiagonal matrix by a certain matrix **Q**, representing the appropriate change of coordinates. (For a detailed study of the similarity transformation, see Section 5.9.)

There is no general procedure for finding the matrix **Q**, although in most cases simple geometrical arguments will suffice. The elements κ_i contain the maximum and minimum values of the susceptibility together with some intermediate value. These three values are observed in three mutually perpendicular directions which are the principal magnetic axes.

7.2 Isotropic susceptibility

Total isotropy is found in the cubic system, characterized by four elements of symmetry C_3 and their corresponding operators **3** and $\mathbf{3}^2$. Magnetic isotropy must be expected in view of symmetry considerations, as follows.

Since the 3-fold rotations always exchange x, y, and z, they must exchange the values of κ_1, κ_2, and κ_3. Since such rotations are symmetry operations, they cannot affect the magnetization. Therefore, κ_1, κ_2, and

κ_3 must be equal. In other words,

$$\kappa = 3^{-1} \cdot \kappa \cdot 3 \tag{7-5}$$

Replacing the operators **3** and $\mathbf{3}^2$ by the matrix notation used in Section 5.13, one obtains

$$\begin{bmatrix} \kappa_2 & 0 & 0 \\ 0 & \kappa_3 & 0 \\ 0 & 0 & \kappa_1 \end{bmatrix} = \begin{bmatrix} 0 & 1 & 0 \\ 0 & 0 & 1 \\ 1 & 0 & 0 \end{bmatrix} \cdot \begin{bmatrix} \kappa_1 & 0 & 0 \\ 0 & \kappa_2 & 0 \\ 0 & 0 & \kappa_3 \end{bmatrix} \cdot \begin{bmatrix} 0 & 0 & 1 \\ 1 & 0 & 0 \\ 0 & 1 & 0 \end{bmatrix} \tag{7-6}$$

In view of Eq. (7-6), Eq. (7-5) can be true only if

$$\kappa_1 = \kappa_2 = \kappa_3 = \kappa \tag{7-7}$$

and the susceptibility is then a scalar, since

$$\begin{bmatrix} \kappa & 0 & 0 \\ 0 & \kappa & 0 \\ 0 & 0 & \kappa \end{bmatrix} = \kappa \begin{bmatrix} 1 & 0 & 0 \\ 0 & 1 & 0 \\ 0 & 0 & 1 \end{bmatrix} = \kappa \tag{7-8}$$

Notice that the similarity transformation by a symmetry operator has been performed on κ since κ is a matrix. Details of the operation performed in Eq. (7-5) and the matrices used in Eq. (7-6) may be found in Sections 5.9 and 5.13.

7.3   The axially symmetric case

The equivalence among x, y, and z is partially removed by distortions of the cubic lattice that transform it into tetragonal, rhombohedral, and hexagonal lattices. Among them, the tetragonal case is the simplest to deal with, since the elements C_4 and S_4, characteristic of such a system, generate operations that exchange x and y, but not z. One may use a symmetry argument analogous to that used in the preceding section and write

$$\kappa = 4^{-1} \cdot \kappa \cdot 4 \tag{7-9}$$

i.e.,

$$\begin{bmatrix} \kappa_2 & 0 & 0 \\ 0 & \kappa_1 & 0 \\ 0 & 0 & \kappa_3 \end{bmatrix} = \begin{bmatrix} 0 & 1 & 0 \\ \bar{1} & 0 & 0 \\ 0 & 0 & 1 \end{bmatrix} \cdot \begin{bmatrix} \kappa_1 & 0 & 0 \\ 0 & \kappa_2 & 0 \\ 0 & 0 & \kappa_3 \end{bmatrix} \cdot \begin{bmatrix} 0 & \bar{1} & 0 \\ 1 & 0 & 0 \\ 0 & 0 & 1 \end{bmatrix} \tag{7-10}$$

Eq. (7-9) is valid, in view of eq. (7-10), only if $\kappa_1 = \kappa_2$, while κ_3 may have any value. The matrix κ, then, represents a cylindrically (axially) symmetric tensor. The notation in such a case is simplified to $\kappa_1 = \kappa_2 = \kappa_\perp$ and $\kappa_3 = \kappa_\parallel$. The matrix then becomes

$$\kappa = \begin{bmatrix} \kappa_\perp & 0 & 0 \\ 0 & \kappa_\perp & 0 \\ 0 & 0 & \kappa_\parallel \end{bmatrix} \qquad (7\text{-}11)$$

That the matrix κ must also be axially symmetric in the rhombohedral and hexagonal cases is not so obvious. However, if κ_3 is now the value of the susceptibility along the direction of the 3-fold and 6-fold axes, it is clear that if one initially lines up the single crystal in the canonical orientation of κ_1 ($H_y = H_z = 0$), and then rotates it by $2\pi/3$ or by $\pi/3$, in each case the magnetization cannot change, for such rotations are operations of symmetry. In the new orientations, however, the susceptibility will have contributions due to κ_1 and κ_2, since the orientations of κ_1 and κ_2 (principal magnetic axes) are perpendicular to each other. Therefore, κ_1 and κ_2 must be equal, i.e.,

$$\kappa_1 = \kappa_2 = \kappa_\perp \qquad (7\text{-}12)$$

The following similarity transformation may be performed by a 3-fold operator about the z-axis

$$\kappa = 3^{-1} \cdot \kappa \cdot 3 \qquad (7\text{-}13)$$

with

$$3 = \begin{bmatrix} \cos 2\pi/3 & \sin 2\pi/3 & 0 \\ -\sin 2\pi/3 & \cos 2\pi/3 & 0 \\ 0 & 0 & 1 \end{bmatrix} \qquad (7\text{-}14)$$

and

$$3^{-1} = \begin{bmatrix} \cos 2\pi/3 & -\sin 2\pi/3 & 0 \\ \sin 2\pi/3 & \cos 2\pi/3 & 0 \\ 0 & 0 & 1 \end{bmatrix} \qquad (7\text{-}15)$$

representing clockwise rotations by $2\pi/3$ and $4\pi/3$ about the z-axis. It should be verified that Eq. (7-13) can only be true if $\kappa_1 = \kappa_2$, while no restriction is imposed upon the value of κ_3.

We then see that tetragonal, rhombohedral, and hexagonal lattices are characterized by axially symmetric tensors of magnetic susceptibility. Results of measurements on polycrystalline samples and randomly oriented powdered crystals yield the average value

$$\langle \kappa \rangle = \frac{1}{3} (\kappa_\parallel + 2\kappa_\perp) \qquad (7\text{-}16)$$

7.4 The totally asymmetric case

The remaining crystalline systems, orthorhombic, monoclinic, and triclinic, may have 2-fold axes, symmetry center, and pinacoidal planes of symmetry, i.e., symmetry planes of Miller indices (100), (010), and (001). All these elements have a common feature: they generate symmetry operations that never exchange x, y, or z. Under such circumstances, it is clear that there are no symmetry arguments imposing restrictions on the values

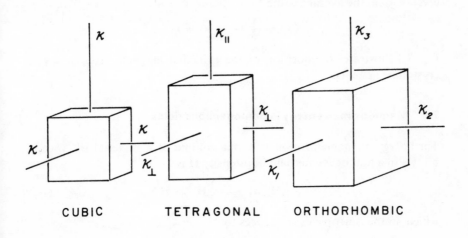

CUBIC   TETRAGONAL   ORTHORHOMBIC

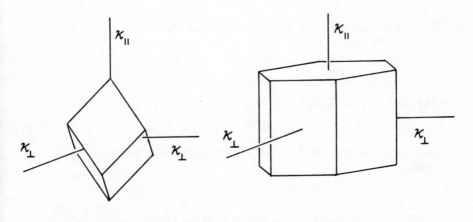

RHOMBOHEDRAL   HEXAGONAL

FIG. 7-1 Orientation of magnetic axes in different crystal systems.

of the principal susceptibilities, since any symmetry operator $\mathbf{R}$ will leave the κ-matrix unchanged.

$$\begin{bmatrix} \kappa_1 & 0 & 0 \\ 0 & \kappa_2 & 0 \\ 0 & 0 & \kappa_3 \end{bmatrix} = \mathbf{R}^{-1} \cdot \begin{bmatrix} \kappa_1 & 0 & 0 \\ 0 & \kappa_2 & 0 \\ 0 & 0 & \kappa_3 \end{bmatrix} \cdot \mathbf{R} \qquad (7\text{-}17)$$

The magnetic tensor is therefore adequately described by Eq. (7-4).

Results on polycrystalline samples and randomly oriented powdered crystals yield the average value

$$\langle \kappa \rangle = \frac{1}{3} (\kappa_1 + \kappa_2 + \kappa_3) \qquad (7\text{-}18)$$

Fig. 7.1 illustrates the position of the principal magnetic axes in several crystalline systems.

7.5 Magnetization energy in homogeneous fields

The energy of magnetization per unit volume of a material of susceptibility κ in a homogeneous field of intensity $\mathbf{H}$ is

$$-\frac{1}{2} \mathbf{H} \cdot \mathbf{M} = -\frac{1}{2} \mathbf{H} \cdot \kappa \cdot \mathbf{H} \qquad (7\text{-}19)$$

which, in the isotropic case, reduces to

$$-\frac{1}{2} \mathbf{H} \cdot \mathbf{M} = -\frac{1}{2} \kappa H^2 \qquad (7\text{-}20)$$

If $\mathbf{i}$, $\mathbf{j}$, $\mathbf{k}$ are the unit vectors of the reference frame that diagonalizes the susceptibility matrix κ (unit vectors in the directions of the principal magnetic axes), vector $\mathbf{H}$ can be written as

$$\mathbf{H} = H_x \mathbf{i} + H_y \mathbf{j} + H_z \mathbf{k} \qquad (7\text{-}21)$$

The product $\kappa \cdot \mathbf{H}$ is therefore

$$\mathbf{M} = \kappa_1 H_x \mathbf{i} + \kappa_2 H_y \mathbf{j} + \kappa_3 H_z \mathbf{k} \qquad (7\text{-}22)$$

and the scalar product by $\frac{1}{2}\mathbf{H}$ that gives the magnetization energy becomes

$$-\frac{1}{2} (\kappa_1 H_x^2 + \kappa_2 H_y^2 + \kappa_3 H_z^2) \qquad (7\text{-}23)$$

Since homogeneity of the magnetic field has been assumed, there are no translational forces applied to the sample, for the derivative

$$-\frac{1}{2} \frac{d}{ds} (\kappa_1 H_x^2 + \kappa_2 H_y^2 + \kappa_3 H_z^2) \qquad (7\text{-}24)$$

with respect to any displacement ds is identically zero. There may be torques, however, for a rotation by $d\Theta$ changes the values of H_x, H_y, and H_z, though always satisfying the condition

$$H^2 = H_x^2 + H_y^2 + H_z^2 \qquad (7\text{-}25)$$

If the substance is anisotropic, the derivative

$$-\frac{1}{2}\frac{d}{d\theta}(\kappa_1 H_x^2 + \kappa_2 H_y^2 + \kappa_3 H_z^2) \qquad (7\text{-}26)$$

which expresses a torque, is not necessarily zero. Eq. (7-26) may be simplified by replacing H^2 and the direction cosines

$$l = H_x/H, \qquad m = H_y/H, \qquad n = H_z/H \qquad (7\text{-}27)$$

for H_x, H_y, and H_z, obtaining

$$-\frac{1}{2}\frac{d}{d\theta}(\kappa_1 l^2 + \kappa_2 m^2 + \kappa_3 n^2)H^2$$

$$= -H^2[\kappa_1 l(dl/d\theta) + \kappa_2 m(dm/d\theta) + \kappa_3 n(dn/d\theta)] \qquad (7\text{-}28)$$

7.6 Determination of the principal axes when one of them is known

Let us assume that we know the direction of the principal axis κ_3. We can cut a circular disc of the crystal with the plane of the disc perpendicular to κ_3 and suspend it with this plane horizontal and parallel to a homogeneous field **H**. The torque will be given by Eq. (7-28) with $n = 0$, i.e.,

$$-\frac{1}{2}\frac{d}{d\theta}(\kappa_1 \cos^2\theta + \kappa_2 \sin^2\theta)H^2 V =$$

$$(\kappa_1 \cos\theta \cdot \sin\theta - \kappa_2 \sin\theta \cdot \cos\theta)H^2 V = \frac{1}{2}(\kappa_1 - \kappa_2)\sin 2\theta \cdot H^2 V \qquad (7\text{-}29)$$

where V is the volume of the crystal and θ the angle between the directions of κ_1 and **H**. The disc will be acted by a torque given by Eq. (7-29), which is clearly a maximum for $\theta = 45°$.

If the disc is suspended in a torsion device permitting the measurement of the torque exerted on the crystal, as well as allowing rotations to modify the value of θ, it is possible to determine the position of the remaining magnetic axes. In addition, if both the field intensity and the crystal volume are known, the difference $(\kappa_1 - \kappa_2)$ may be measured. This procedure is of special value for studying monoclinic crystals, in which one of the principal axes can always be found if the orientation of the crystal is known. Several methods permit to determine the orientation: direct observation before cutting, X-ray diffraction, polarizing microscopy, chemical etching, etc. The applicability of the torsion method is limited by the magnitude $(\kappa_1 - \kappa_2)$, since the torque measured is proportional to this difference. This method, to which we will refer as the *Static Torsion Method*, leads to an interesting variation—the *Dynamic Torsion Method*, which consists of measuring the oscillation period of the disc suspended between the poles of an electromagnet in the presence and in the absence of a magnetic field.[6] Assume that the disc is suspended as in the preceding method, but that the position of all three principal magnetic axes is already known (by the preceding method). The disc will oscillate, in the absence of a magnetic field, with a certain period T_o, due to a restoring torque

$$I(d^2\theta/dt^2) = -C\delta\theta \tag{7-30}$$

where C is the torsion constant of the suspension fiber, I the moment of inertia of the disc, and $\delta\theta$ a small angular deviation from the equilibrium position. If a magnetic field is applied in the direction of κ_1, the general expression for the additional restoring torque will be given by Eq. (7-29). For the small deviation $\delta\theta$, the following approximation holds:

$$\sin 2\delta\theta \approx 2\delta\theta \tag{7-31}$$

The expression for the additional torque may then be simplified to

$$(\kappa_1 - \kappa_2)H^2V\delta\theta \tag{7-32}$$

The total restoring torque will then be

$$I(d^2\theta/dt^2) = -[C + (\kappa_1 - \kappa_2)H^2V]\delta\theta \tag{7-33}$$

and the oscillation period in the presence of the magnetic field, T, will be related to T_o by

$$CT_o^2 = [C + (\kappa_1 - \kappa_2)H^2V]T^2 \tag{7-34}$$

so that

$$\kappa_1 - \kappa_2 = \frac{T_o^2 - T^2}{T^2} \cdot \frac{C}{H^2} \cdot \frac{1}{V} \tag{7-35}$$

The oscillation method has been successfully applied to the study of diamagnetic anisotropy in organic crystals.[6] The accuracy of the oscillation method, as the static case, is limited by the difference $(\kappa_1 - \kappa_2)$. When this difference is very small, the intrinsic sensitivity of the device may not be large enough and any asymmetry of shape may affect the results. In order to minimize this error, measurements have been made having the sample immersed in a solution whose volume susceptibility was equal to that of the powdered crystals.[5]

7.7 Magnetization in inhomogeneous fields

When a sample is placed in a nonuniform field, Eq. (7-23) is valid in each point, provided that the components of the magnetic field be replaced by the corresponding functions of the coordinates. The magnetic energy of a sample of volume V is then given by the integral

$$E = -\frac{1}{2} \int_V (\kappa_1 H_x^2 + \kappa_2 H_y^2 + \kappa_3 H_z^2) dV \qquad (7\text{-}36)$$

By using the direction cosines introduced in Eqs. (7-27) and

$$\kappa_1 l^2 + \kappa_2 m^2 + \kappa_3 n^2 = \kappa_\theta \qquad (7\text{-}37)$$

where the subscript θ is used for symbolizing the observed value of κ along the direction of the magnetic field, one may rewrite Eq. (7-36) as

$$E = -\frac{1}{2} \int_V \kappa_\theta H^2 dV \qquad (7\text{-}38)$$

Since it is assumed that

$$\mathbf{H} = \mathbf{H}(x, y, z) \qquad (7\text{-}39)$$

the derivative shown in Eq. (7-26) is no longer zero, but

$$-\frac{1}{2} \frac{d}{ds} (\kappa_1 H_x^2 + \kappa_2 H_y^2 + \kappa_3 H_z^2)$$
$$= -[\kappa_1 H_x (dH_x/ds) + \kappa_2 H_y (dH_y/ds) + \kappa_3 H_z (dH_z/ds)] \quad (7\text{-}40)$$

7.8 Determination of the principal susceptibilities when the axes are known

By considering Eq. (7-40) it follows that, if the principal magnetic axes are known, one may apply a field with the gradient parallel to each axis in turn and measure the principal magnetic susceptibilities directly.[4] In most cases, the crystals are small; they are cut to a spherical shape in order to avoid spurious volume effects and weighed using the Faraday technique. On exciting the electromagnet, the apparent change in weight $g\delta m$ is

$$g\delta m = \kappa_i V H (dH/ds) \qquad (7\text{-}41)$$

where κ_i is the principal susceptibility in each case parallel to the magnetic field, and s the vertical coordinate. Notice that in the conventional methods the principal susceptibility which is measured is *not* the one to which the gradient is parallel.

7.9 Measurement on monoclinic rods

An effect that has not been considered so far is the influence of orientation in measurements using Gouy's method. This effect has been used to determine two principal axes in cylindrical single crystals of metals and semiconductors with the position of the third axis known.[1,2,3] This method has recently increased in importance due to developments in the techniques for growing and purifying long metallic rods, each formed of a single crystal. Although a crystallographic axis does not generally coincide with the axis of the rod, such coincidence will first be assumed to avoid unnecessary complications adding nothing to the understanding of the effect.

If the rod in a Gouy device is suspended with its long axis vertical (i.e., with κ_3 perpendicular to the lines of force), the total magnetic pull will depend on the orientation of κ_1 and κ_2 with respect to H, provided that κ_1 and κ_2 are different. The force acting upon any element of the rod, of length dz, is given by

$$dF = \kappa_\theta A H \,(dH/dz)\, dz \tag{7-42}$$

where A is the cross section of the rod and κ_θ is the susceptibility in the direction of H, i.e.,

$$\kappa_\theta = \kappa_1 \cos^2\theta + \kappa_2 \sin^2\theta = \kappa_2 + (\kappa_1 - \kappa_2)\cos^2\theta \tag{7-43}$$

where, as in Eq. (7-29), θ is the angle between the directions of H and κ_1. Introduction of Eq. (7-43) in (7-42) and integration between H at $z = 0$ and H_L at $z = L$, the length of the rod, leads to the total pull

$$F_z = [\kappa_2 + (\kappa_1 - \kappa_2)\cos^2\theta](A/2)(H^2 - H_L^2) \tag{7-44}$$

As usual, if the length of the rod, which is supposed to be of constant cross section, is large in respect to the diameter of the magnet pole tips, then, $H_L \ll H$, and the term H_L^2 may be neglected. If the total pull is counterbalanced by the weight $g\delta m$, one may write

$$g\delta m = [\kappa_2 + (\kappa_1 - \kappa_2)\cos^2\theta](A/2)\,H^2 = [\chi_2 + (\chi_1 - \chi_2)\cos^2\theta](A/2)\,H^2\rho \tag{7-45}$$

where χ_i is the mass susceptibility and ρ the density of the rod.

In order to orient conveniently the rod for determining the directions of κ_1 and κ_2, it must be held in position by a rigid suspension allowing rotation about the direction of κ_3. Once the directions of κ_1 and κ_2 have been determined, their values are given by

$$g\delta m_1 = \kappa_1(A/2)\,H^2 \quad \text{at} \quad \theta = 0° \tag{7-46}$$

and

$$g\delta m_2 = \kappa_2(A/2)H^2 \qquad \text{at} \qquad \theta = 90° \qquad (7\text{-}47)$$

The same comments of Section 7.6 about shape asymmetries stand in this case.

When the rod axis does not contain one principal axis of susceptibility, the problem becomes very involved and independent measurements are needed with different orientation of the principal axes of susceptibility with respect to the long axis of the rod. The problem is somewhat simplified in the case of magnetic axial (cylindrical) symmetry of principal values $\kappa_\parallel$ and $\kappa_\perp$.[7] In this case, let OP in Fig. 7.2 be the direction of $\kappa_\parallel$, making an angle ϕ with the rod long axis, and OR be the intersection

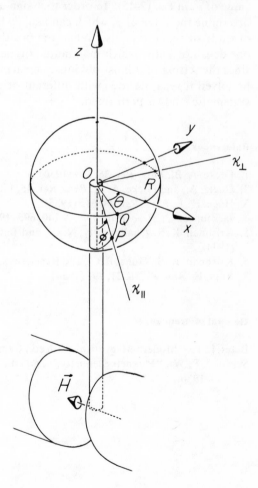

FIG. 7-2 Determination of the principal mangetic susceptibilities of a single-crystal rod.

of the plane containing $\kappa_\perp$ with the plane of the cross section of the rod. Let x be the direction of the lines of force and z that of the vertical. The susceptibility along OQ is

$$\kappa_\phi = \kappa_\parallel \sin^2 \phi + \kappa_\perp \cos^2 \phi \qquad (7\text{-}48)$$

and the susceptibility along x is then

$$\kappa_\theta = (\kappa_\parallel \sin^2 \phi + \kappa_\perp \cos^2 \phi) \cos^2 \theta + \kappa_\perp \sin^2 \theta \qquad (7\text{-}49)$$

where ϕ is the angle between OP and z, and θ the angle between OQ and the magnetic vector (i.e., the x-axis).

The function of Eq. (7-49) has two maxima: at $\theta = 0°$, and at $\theta = 90°$. That at $\theta = 90°$ gives the value of $\kappa_\perp$ alone. That at $\theta = 0°$ gives the value of κ_ϕ in Eq. (7-45). In order to assign a value to $\kappa_\parallel$ it is necessary to determine the value of ϕ, which can usually be made by standard methods of study of single crystals. When the position of the axes is not known, one does not know which maximum corresponds to $\kappa_\perp$ and which to κ_ϕ, since the setting of θ must obviously be arbitrary. The problem can only be solved if separate rods with different orientations are available, and a systematic study is performed.

References

1. Focke, A. B., *Phys. Rev.* **36**, 319 (1930).
2. Goetz, A., and Focke, A. B., *Phys. Rev.* **45**, 170 (1934).
3. Hoge, H. J., *Phys. Rev.* **48**, 615 (1935).
4. Jackson, L. C., *Proc. Roy. Soc.,* A **140**, 695 (1933).
5. Krishnan, K. S., Chakravorty, N. C., and Banerjee, S., *Phil. Trans.,* A **232**, 99 (1933).
6. Krishnan, K. S., Guha, B. C., and Banerjee, S., *Phil. Trans.,* A **231**, 235 (1933).
7. Vogt, E., *Ann. Physik* **21**, 791 (1934).

General references

Bates, L. F., "Modern Magnetism," 4th ed., Cambridge, University Press, 1961.
Selwood, P. W., "Magnetochemistry," 2nd ed., New York, Interscience Publishers, 1956.

PART **III**

Electron paramagnetic resonance

8 *Experimental methods*

8.1 Electron paramagnetic resonance: an introductory outline

Electron paramagnetic resonance (EPR), paramagnetic resonance (PR), or electron spin resonance (ESR) (as it is called by different workers) was discovered by Zavoisky[24] in the Soviet Union in 1945 and, independently, a year later by Cumerow and Halliday[8] in the United States. Although the three names, EPR, PR, and ESR are indistinctly used, EPR seems to be more appropriate. Paramagnetic resonance, the name preferred by the Oxford workers, could cover resonance of any type of permanent magnetic moments—including the nuclear ones, for which the name nuclear magnetic resonance (NMR) is specifically used. Electron spin resonance conveys the erroneous idea that only the electron spin plays a significant role, while ignoring the important orbital contribution. Electron paramagnetic resonance eliminates ambiguities and therefore should be preferred.

In order to develop the basic principle of operation of EPR, it is advantageous to start with a simple case—namely, a substance having atoms, molecules or ions with one unpaired electron, with the orbital angular momentum almost completely quenched (Section 2.8). These paramagnetic units are further supposed to be diluted in a diamagnetic environment; for in such a case neighboring unpaired electrons will be sufficiently apart to involve no appreciable interaction. In the absence of a magnetic field, the state of the electrons will be two-fold spin degenerate (Kramer's theorem, Section 2.7); all the unpaired electrons may then be described as having the same energy E_o indicated in Fig. 3.1. Strictly speaking, there will be an energy distribution about E_o, following Maxwell-Boltzmann's statistics, *if there is no spin-spin interaction,* according to what was seen in Section 3.3. For the purpose of discussing electron paramagnetic resonance however, it will be sufficient to consider all the electrons as having the mean thermal energy plus averaged contributions arising from non-magnetic interactions. Application of a magnetic field will remove the spin degeneracy, according to Kramer's theorem, and the energy level E_o will be split into two levels of energies

$$E_i = E_o + m_s g \mu_B H \qquad (8\text{-}1)$$

with $m_s = \pm \frac{1}{2}$, illustrated in Fig. 3.1. The same comment about average thermal and non-magnetic energy contributions stands in this case. The state of lower energy corresponds to parallel alignment of the electron magnetic moment in the external magnetic field. In writing Eq. (8-1), the magnetic term $m_s g \mu_B H$ has been added. Parallel alignment of the magnetic moment of the electron with the magnetic field, which has the lowest energy, then corresponds to $m_s = -\frac{1}{2}$, and m_s represents the quantized projection of the S-vector (divided by $\hbar$) along **H**. The magnetic moment associated with S has an opposite sign, as seen in Section 2.1. Eq. (8-1) could have been written by *subtracting* the magnetic term $m_s g \mu_B H$. Had this been the case, m_s would have represented the quantized magnetic moment along **H**, which must be positive for parallel alignment. It is certainly a matter of taste to choose either alternative, although the use of quantum mechanics has imposed addition of the magnetic term as in Eq. (8-1), for the quantity that appears in a natural way is the spin vector operator. We shall, then, add the magnetic term and interpret m_s as the quantized projection of $S/\hbar$ along **H**.

The relative population of both levels clearly depends on the temperature T. If the electrons are in thermal equilibrium with the lattice, the population ratio N_2/N_1 is given by Maxwell-Boltzmann's statistics as

$$N_2/N_1 = exp[-(E_2 - E_1)/kT] \qquad (8\text{-}2)$$

where k is Boltzmann's constant. Since the net magnetization of the sample is proportional to $(N_1 - N_2)$, it follows that it will increase with decreasing temperature and will tend to vanish at a sufficiently high temperature. It has been seen in Section 3.3 that the linear approximation of Eq. (8-2) leads to Curie's law.

Irradiation of the sample with photons of energy equal to the difference

$$E_2 - E_1 = \Delta E = g\mu_B H = h\nu \tag{8-3}$$

will cause electrons in the lower level to absorb energy and be raised to the upper level. Since promotion of electrons from E_1 to E_2 is counterbalanced by their interaction with the lattice (spin-lattice interaction), through which they return to the lower level by a radiationless process, a steady absorption of photons may take place. Steady absorption of photons then results in lattice heating. At the same time, a finite spin-lattice relaxation time, which increases with decreasing temperature, limits the power fed into the sample; otherwise, saturation would occur. In other words, if the power fed exceeds certain limits, the rate at which electrons are raised becomes greater than the rate at which they can return to the lower level, both levels soon are equally populated, and absorption of power practically ends.

Measurement of the frequency ν of the photons and of the value of the magnetic field will suffice to determine the actual value of g, which is one of the objectives of EPR, since

$$h\nu = g\mu_B H \tag{8-4}$$

Replacing the figures for the constants, it is found, for example, that for magnetic fields of the order of three kilo-oersted, the frequency should be of the order of some nine Gigacycle/second if the g-factor is close to two. This frequency corresponds to a wave length of 3 cm (in the microwave region) and must be generated by klystrons and transmitted by wave guides of certain geometrical characteristics. Since each g-value actually determines the value of the ratio ν/H, it is clear that lower frequencies and fields could also be used, in which case conventional LC circuits would suffice. There are, however, definite advantages in using the more elaborate microwave technique, among them significantly higher sensitivity, and magnetic fields much stronger than the field at the electron due to surrounding nuclei of magnetic moments other than zero.

8.2 Basis of the experimental detection of resonance

When microwave frequencies are used in EPR experiments, the condition of resonance must be searched for by varying the magnetic field, because microwave sources can only be tuned to within a very narrow range

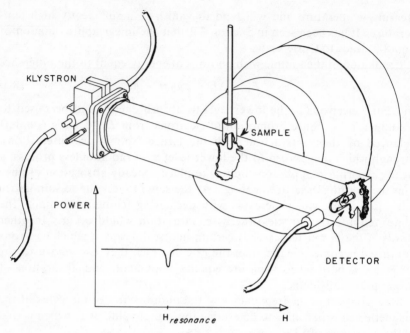

FIG. 8-1 Microwave transmission spectrometer.

of frequencies, usually of the order of a few percent. A simplified scheme
of transmission spectrometer is illustrated in Fig. 8.1. A klystron generates microwaves that are transmitted down a wave guide of rectangular
cross section and metallic walls. The microwave power transmitted down
the line is detected by a crystal rectifier chosen in such a way that the crystal current is proportional to the microwave power falling on it. Some
absorbing material (usually graphite-coated plastic) prevents reflections in
the wall behind the detector. A paramagnetic sample is placed just inside
the wave guide through a hole in the narrow face. The static magnetic
field generated by the electromagnet (one of whose pole pieces is visible
in the figure) is made to vary at a slow rate as the crystal current is plotted
against H. A constant level of microwave power falling on the detector
is thus recorded until the value of the resonance field is reached. At this
point, absorption by the sample increases, thus decreasing the microwave
power reaching the detector, Absorption by the sample is then detected as
a decrease in the crystal current, shown in the recording at the bottom of
the figure. As soon as the condition of resonance is no longer fulfilled, the
power falling on the crystal increases, and the crystal current rises to its
former value. Although conceptually simple, this device is not a sensitive
one. Other devices, based on a special array of transmission lines permit-

ting the system to balance in the absence of resonance, and a resonant cavity where standing waves are formed, provide extremely large sensitivities to the extent of making possible the detection of 10^{11} spins for a line width of one oersted. But before attacking the study of such systems, it is necessary to elaborate a little more about the theory of the resonance itself.

8.3 Complex susceptibilities

The susceptibility of a paramagnetic substance was derived in Section 3.3 assuming static conditions. Its measurement was analogously considered in Chapters 6 and 7 in a d.c. magnetic field, or in an a.c. magnetic field of moderately low frequency (quasistatic electromagnetic methods). The case of electron paramagnetic resonance poses a different problem. At the high frequencies commonly used in EPR experiments (ten and more Gc/s), the magnetization of the specimen does not follow the variation of the oscillating magnetic field instantaneously, thereby showing a phase lag. The general expression for the susceptibility must then be written

$$\kappa = \kappa' - j\kappa'' \tag{8-5}$$

where κ', the real part of the susceptibility, accounts for the in-phase magnetization, while the imaginary part, κ'', determines the out-of-phase magnetization. The imaginary number $j = \sqrt{-1}$ indicates, as usual, a phase difference of $\pi/2$. The total susceptibility is then given, in absolute value, by

$$\kappa = (\kappa'^2 + \kappa''^2)^{1/2} \tag{8-6}$$

If the microwave magnetic field is

$$H_1 \sin \omega t \tag{8-7}$$

where $\omega = 2\pi\nu$, it will determine an in-phase magnetization

$$M' = \kappa' H_1 \sin \omega t \tag{8-8}$$

and an out-of-phase magnetization

$$M'' = -\kappa'' H_1 \cos \omega t \tag{8-9}$$

where the phase shift by $\pi/2$ is accounted for by substituting $\cos \omega t$ for $\sin \omega t$.

Since we are interested in absorption of microwave power by the sample, we have to write the rate at which the magnetic energy changes per unit time, i.e.,

$$H(dM/dt) = H d(M' + M'')/dt \tag{8-10}$$

The total power is clearly obtained by integrating Eq. (8-10) over one cycle and dividing by the oscillation period

$$T_{osc} = 1/\nu = 2\pi/\omega \tag{8-11}$$

namely:

$$P = (\omega/2\pi) \int_0^{2\pi/\omega} H\,(dM/dt)\,dt$$

$$= (\omega/2\pi)\,H_1^2 \int_0^{2\pi/\omega} (\sin \omega t \cdot \kappa' \cos \omega t + \kappa'' \sin^2 \omega t)\,dt \tag{8-12}$$

which follows after differentiation of Eqs. (8-8) and (8-9) and substitution in Eq. (8-10). The integral of Eq. (8-12) has two terms, i.e.,

$$\frac{\kappa'\omega^2}{2\pi} \int_0^{2\pi/\omega} (\sin \omega t \cdot \cos \omega t)\,dt \tag{8-13}$$

and

$$\frac{\kappa''\omega^2}{2\pi} \int_0^{2\pi/\omega} \sin^2 \omega t\,dt \tag{8-14}$$

Eq. (8-13) can be written as

$$\frac{\omega\kappa'}{2\pi} \int_0^{2\pi} \sin \omega t \cdot \cos \omega t\,d(\omega t) = 0 \tag{8-15}$$

It follows that in-phase magnetization does not absorb microwave power. On the other hand, Eq. (8-14) may readily be integrated to

$$\frac{\omega^2\kappa''}{2\pi} \int_0^{2\pi/\omega} \sin^2 \omega t\,dt = \frac{\omega\kappa''}{2\pi} \int_0^{2\pi} \sin^2 \omega t\,d(\omega t) = \frac{\omega\kappa''}{\pi} \int_0^{\pi} \sin^2 x\,dx = \frac{\omega\kappa''}{2} \tag{8-16}$$

Eq. (8-16) accounts for absorption of microwave power. Introduction of Eqs. (8-15) and (8-16) into Eq. (8-12) leads to the total power absorbed by the sample

$$P = \frac{1}{2}\,\omega\kappa''\,H_1^2 \tag{8-17}$$

Therefore, the energy absorbed per cycle is

$$W_C = P\,(2\pi/\omega) = \pi\kappa''\,H_1^2 \tag{8-18}$$

Effects due to in-phase and out-of-phase magnetization are called *paramagnetic dispersion* and *paramagnetic absorption*; they may be independently observed in EPR experiments.

8.4 Absorption of power by electronic transitions

The number of energy levels in a magnetic field of an electron of total quantum number J is $2J + 1$, corresponding to $m_J = -J, -(J - 1), \ldots, +J$. These levels are equally spaced, the energy difference being, in each case, $g\mu_B H$, i.e., for any transition characterized by $\Delta m_J = \pm 1$. For ordinary values of H and not very low temperatures, the difference in population between the levels is given by the linear approximation of Boltzmann's distribution, so that the level characterized by the resolved component m_J will have a population

$$\left[1 + \frac{m_J g\mu_B H}{kT} \right] \frac{N_0}{2J + 1} \tag{8-19}$$

where N_0 is the total number of unpaired electrons per unit volume. Clearly, summation of Eq. (8-19) over the $2J + 1$ levels leads to N_0, since

$$\sum_{m_J = -J}^{+J} \left[1 + \frac{m_J g\mu_B H}{kT} \right] = 2J + 1 \tag{8-20}$$

The coefficient $N_0/(2J + 1)$ of Eq. (8-19) is then a normalization factor.

Since absorption of microwave power occurs because electrons are promoted to higher levels, the power absorption evaluated as a function of the imaginary part of the susceptibility in Eq. (8-17) must be proportional to the difference in population between adjacent levels and to the transition probability. The difference in population between adjacent levels of m_J and $(m_J - 1)$ follows from Eq. (8-19)

$$\frac{N_0}{2J + 1} \left[1 + \frac{m_J g\mu_B H}{kT} - 1 - \frac{(m_J - 1)g\mu_B H}{kT} \right] \tag{8-21}$$

If we introduce the condition of resonance

$$g\mu_B H = h\nu_0 \tag{8-4}$$

the expression for the difference in population given in Eq. (8-21) reduces to

$$\frac{N_0 h\nu_0}{(2J + 1)kT} \tag{8-22}$$

Each transition of an electron from a lower level to the upper adjacent one occurs with an energy absorption of $h\nu$ close or equal to $h\nu_0$, accounting for a certain energy distribution of the electrons about each level. If $p(m_J)$ is the upward transition probability from level $(m_J - 1)$ to level

m_J, the net absorption of power between levels is

$$P(m_J) = \frac{N_0 h^2 \nu \nu_0}{(2J+1)kT} p(m_J) = \frac{N_0 \hbar^2 \omega \omega_0}{(2J+1)kT} p(m_J) \tag{8-23}$$

where $\hbar = h/2\pi$, and ω is the angular frequency in radians per second.

If $f(\omega - \omega_0)$ is the *line shape function* that represents the shape of the absorption line, standard radiation theory gives for the transition probability the expression

$$p(m_J) = (\pi/4)\gamma^2 H_1^2 (J + m_J)(J - m_J + 1) f(\omega - \omega_0) \tag{8-24}$$

where

$$\gamma = g\mu_B/h \tag{8-25}$$

is an alternative expression of the magnetogyric ratio of the electron introduced in Section 1.4. In order to prove it, it is necessary to remember that the absorption of energy accompanying a transition of Δm_J is

$$g\mu_B H \Delta m_J = H (\Delta m)_H \tag{8-26}$$

where $(\Delta m)_H$ is the change in the component of the magnetic moment in the direction of **H**. If $(\Delta G)_H$ is the change in the component of the angular momentum along **H**, it follows, from Eq. (1-28),

$$(\Delta m)_H = \gamma (\Delta G)_H \tag{8-27}$$

where

$$(\Delta G)_H = \hbar \Delta m_J \tag{8-28}$$

It can then be writen

$$g\mu_B H \Delta m_J = H(\Delta m)_H = H\gamma(\Delta G)_H = H\gamma\hbar\Delta m_J \tag{8-29}$$

with the expression of γ given by Eq. (8-25).

The line shape function $f(\omega - \omega_0)$ is normalized by making

$$\int_0^\infty f(\omega - \omega_0) d\omega = 1 \tag{8-30}$$

In most cases it approximates a Gaussian or Lorentzian function according to the type of broadening that prevails (See Section 8.9 and Chapter 11). The maximum value of $f(\omega - \omega_0)$ is inversely proportional to the line width. Clearly, the line width may be defined in several ways, among them at maximum slope or at half power.

Returning to the microwave power absorbed by the sample, introduction of Eq. (8-24) in Eq. (8-23) leads to

$$P(m_J) = \frac{N_0 \pi \gamma^2 H_1^2 \hbar^2 \omega \omega_0}{4kT(2J+1)} (J + m_J)(J - m_J + 1) f(\omega - \omega_0) \tag{8-31}$$

The total power associated with the $2J$ permitted transitions among the $(2J + 1)$ levels occurring at the same resonance condition is then

$$P = \frac{N_0 \gamma^2 H_1^2 h^2 \omega \omega_0 \pi}{4kT(2J + 1)} f(\omega - \omega_0) \sum_{m_J = -(J-1)}^{+J} (J^2 - m_J^2 + J + m_J) \qquad (8\text{-}32)$$

The summation of Eq. (8-32) is

$$\sum_{m_J = -(J-1)}^{+J} (J^2 - m_J^2 + J + m_J) = \tfrac{2}{3} J(J + 1)(2J + 1) \qquad (8\text{-}33)$$

By replacing $g^2 \mu_B^2 / \hbar^2$ for γ^2 (Eq. 8-25), introducing Eq. (8-33) in Eq. (8-32) and rearranging variables one arrives at

$$P = \frac{\omega H_1^2}{2} \frac{N_0 g^2 \mu_B^2 J(J + 1)}{3kT} \omega_0 \pi f(\omega - \omega_0) \qquad (8\text{-}34)$$

which must be equal to the total power given by Eq. (8-17). Therefore

$$\kappa'' = \left[\frac{N_0 g^2 \mu_B^2 J(J + 1)}{3kT} \right] \omega_0 \pi f(\omega - \omega_0) \qquad (8\text{-}35)$$

The expression in brackets is precisely equal to the static susceptibility (Eqs. 3-56 and 3-57) per unit volume. Thus

$$\kappa'' = \kappa \omega_0 \pi f(\omega - \omega_0) \qquad (8\text{-}36)$$

which is maximum when the condition of resonance $\omega = \omega_0$ is fulfilled, according to the definition of the line shape function $f(\omega - \omega_0)$, for κ is a constant at each temperature.

8.5 The resonant cavity

In the experiment described in Section 8.2, the sample was in the path of a microwave travelling down a wave guide. The electromagnetic energy per unit volume associated with the travelling microwave is usually very small, and so is the sensitivity of the detection itself. It is possible to increase the electromagnetic energy by several orders of magnitude if a device, where standing microwaves may take place, is used. Such a device may be a cavity fulfilling certain geometrical conditions imposed by the wavelength of the electromagnetic radiation. It is then possible to avail electromagnetic energies 10,000-fold larger and more. One such cavity is shown in Fig. 8-2. It is known as a rectangular $H_{012} = TE_{102}$ resonant cavity. The $H_{01} = TE_{10}$ is the term given to the dominant mode propagated in

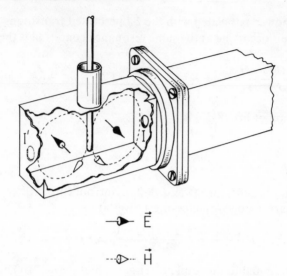

$$\longrightarrow \vec{E}$$

$$\cdots\!\!\!\diamond\!\!\cdots \vec{H}$$

FIG. 8-2 Microwave cavity operating in the TE_{102} mode.

the rectangular guide.* Subscript 2 refers to the existence, as indicated in Fig. 8-2, of two half-wave lengths resonating between the small walls. The power is injected through an iris I bored in one of the small, shorting walls, if the cavity is at the end of the line as shown. Otherwise, an additional iris I' bored in the opposite face provides the necessary coupling if the cavity is placed along a transmission line.

Once the standing waves have been built to their maximum strength, the power fed into the cavity equals the energy loss per unit time. The ability of the cavity to concentrate power is measured by its *Q-factor*, which is defined as

$$Q = \frac{\text{Energy stored}}{\text{Power loss}}\, \omega \qquad (8\text{-}37)$$

Notice that the power loss is the energy dissipated per unit time, which in turn is equal to the energy dissipated per cycle multiplied by the fre-

---

* The nomenclature is as follows: *TE* means transverse electric vector, i.e., perpendicular to the direction of propagation. Subindex 1 means that there is one-half wavelength across the waveguide, in the direction of the narrow side. Subindex 0 indicates absence of electric field in the direction of the broad side. The wavelength of the radiation in the guide, λ_g, differs from the free-space value, λ_{fs}, in that $1/\lambda_{fs}^2 = 1/\lambda_g^2 - 1/\lambda_c^2$ where λ_c is the cut-off wavelength, equal to twice the width of the broad side of the guide if the TE_{10} mode is dominant.

quency, in which case one may write

$$Q = 2\pi \frac{\text{Energy stored in the cavity}}{\text{Energy dissipated per cycle}} \qquad (8\text{-}38)$$

The denominator of Eqs. (8-37) and (8-38) is due, in the absence of sample, to resistive losses in the cavity walls. Upon introduction of the sample, the power loss is naturally increased, even in the absence of resonance. This power loss clearly results in a loss of sensitivity due to a decrease of Q, but may be minimized if the sample is small with respect to the cavity, and if the specific loss of the sample itself is small. Unfortunately, the first solution limits the absolute number of unpaired spins, while the second one is not always feasible.

When the resonance condition is fulfilled, the loss in the sample increases, determining a decrease in the Q of the cavity. This latter effect in turn determines a reduction in the oscillation level, which is usually measured for detecting resonance. An additional effect occurs: in the resonance condition, the characteristics of the *system cavity + sample* change, this change resulting in a shift of the cavity resonant frequency. Clearly, measurement of such shift also provides a method for studying resonance. Both methods of observation may be used. Measurement of the reduction in the oscillatory level permits one to evaluate the absorptive susceptibility component κ''. Measurement of the shift in resonant frequency, on the other hand, allows an evaluation of the dispersive component κ'.

8.6 Cavity-equivalent circuits; absorption and dispersion

The resonant cavity may be represented by the equivalent circuit illustrated in Fig. 8.3. The impedance Z of the circuit is

$$Z = R + j\left(\omega L - \frac{1}{\omega C}\right) \qquad (8\text{-}39)$$

where R represents the resistance of the cavity walls, and L is the inductance. Let us assume, after introducing the sample, that κ represents the change in specific magnetization, at resonance. Then we can write

$$L = L_0(1 + 4\pi\kappa) \qquad (6\text{-}33)$$

according to what was seen in Section 6.7. Since the sample does not fill the cavity completely, Eq. (6-33) must then be written

$$L = L_0(1 + 4\pi\xi\kappa) \qquad (6\text{-}34)$$

where $\xi(0 < \xi \leq 1)$ is the filling factor defined in Section 6.7.

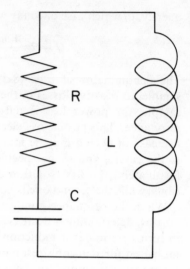

FIG. 8-3   L-C equivalent circuit of a resonating cavity.

Since the susceptibility is given by

$$\kappa = \kappa' - j\kappa'' \tag{8-5}$$

the impedance Z_s at resonance is

$$Z_s = R + j\left[\omega L_0(1 + 4\pi\xi\kappa) - \frac{1}{\omega C}\right]$$

$$= R + 4\pi\xi\omega L_0\kappa'' + j\left[\omega L_0 + 4\pi\xi\omega L_0\kappa' - \frac{1}{\omega C}\right] \tag{8-40}$$

which means that the imaginary component of the susceptibility increases the circuit loss by

$$\Delta R = 4\pi\xi\omega L_0\kappa'' \tag{8-41}$$

while the real component increases the circuit inductance by

$$\Delta L = 4\pi\xi\omega L_0\kappa' \tag{8-42}$$

Let us now consider how the condition of resonance affects the Q of the circuit, defined by

$$Q = \omega L/R \tag{8-43}$$

Notice that the Q of the RLC circuit defined in Eq. (8-43) has the same physical meaning as the Q of the cavity defined in Eq. (8-38), for the storage of energy is represented by L, while R accounts for the power loss.

The total differential of Q is

$$dQ = \frac{\omega}{R} dL - \frac{\omega L}{R^2} dR = \frac{1}{R}(\omega dL - QdR) \tag{8-44}$$

Assuming that the change in susceptibility is small, as indeed it is, we may set

$$dL = 4\pi\xi\omega L_0 \kappa' \tag{8-45}$$

and

$$dR = 4\pi\xi\omega L_0 \kappa'' \tag{8-46}$$

according to Eqs. (8-41) and (8-42). Eq. (8-44) then becomes

$$dQ = \omega Q 4\pi\xi\kappa' - Q^2 4\pi\xi\kappa'' \tag{8-47}$$

Since we learned in Section 8.3 that only the out-of-phase component of the susceptibility accounts for absorption of power (Eq. 8-17), we may write

$$(\partial Q)_L = -Q^2 4\pi\xi\kappa'' \tag{8-48}$$

for the variation of the cavity Q-value that accounts for a decrease in the oscillatory level, to which we referred in Section 8.5.

The in-phase component of Eq. (8-47) accounts for a shift in the cavity resonant frequency. The angular frequency ω_r in the presence of resonance is given by

$$\omega_r L_0(1 + 4\pi\xi\kappa') - \frac{1}{\omega_r C} = 0 \tag{8-49}$$

Thus

$$\omega_r = [L_0(1 + 4\pi\xi\kappa')C]^{-1/2} \tag{8-50}$$

and the change in resonant angular frequency is given by differentiating Eq. (8-49):

$$d\omega_r = -\frac{\omega_r}{2L} dL = -2\pi\xi\omega_r^2\kappa' \tag{8-51}$$

where dL has been substituted by its value given in Eq. (8-45), and $4\pi\xi\kappa'$ neglected in the expression of L in the denominator ($L \approx L_0$). The shift in resonant frequency, as anticipated in the preceding section, is then a *dispersive effect*, since it is accounted for by the in-phase component of the susceptibility.

From the foregoing considerations, it is clear that in EPR experiments one may measure *absorption* (by measuring the change in the oscillatory level inside the cavity) or *dispersion* (by measuring the shift of the resonant frequency). The first approach is usually preferred. For this reason, we will assume throughout the book, unless otherwise stated, that we are measuring absorption.

8.7 Balanced-bridge devices

The simple method of detection described in Section 8.2 (Fig. 8.1) could be significantly improved in absolute sensitivity if a cavity were added in the wave guide, between the poles of the electromagnet, for in such a case the sample could be placed in a region of standing waves of much higher energy. Although the incorporation of the resonant cavity increases the absolute sensitivity of a transmission spectrometer, there still remains the problem of measuring a small variation of a large power level. A definite improvement is obtained with an array of wave guides that permits the system to balance in the absence of resonance, allowing the detector to pick up only the unbalance produced when the condition of resonance is fulfilled, thereby achieving a very high sensitivity. This system essentially consists of a four-arm element that may have the form of a "magic-T" or a "hybrid ring," as illustrated in Fig. 8.4. The resonant cavity is placed at the end of arm 2. The power is fed into arm 1 and is equally divided between arms 2 and 3 if arm 3 matches arm 2 (i.e., if the absorption of microwave power is equal in arms 2 and 3). In such a case, no power reaches the detector, located in arm 4. When paramagnetic

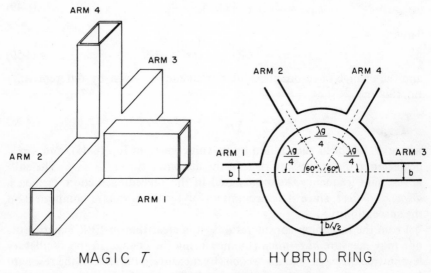

MAGIC T HYBRID RING

FIG. 8-4 Zero-balance microwave devices.

resonance occurs, the absorption in the cavity increases. As a consequence of the power unbalance created between arms 2 and 3, power is fed into arm 4, where it is detected. Although this is the basic principle of operation of the widely used balanced-bridge devices, it should be noted that a small unbalance in the absence of resonance is purposely introduced to obtain maximum sensitivity. On the other hand, if the bridge is initially perfectly balanced, the unbalance produced by the condition of resonance will feed a mixture of dispersion and absorption signals into arm 4. Either of these can be selected if the appropriate amount of unbalance is introduced in the absence of resonance. The magic-T and the hybrid ring differ in that the magic-T obtains its properties from polarization, while the hybrid ring depends on interference effects.

Further discussion of this subject would take us far afield; the interested reader is referred to the specialized literature cited at the end of Part III.

8.8 Field modulation: phase-sensitive detection

In most cases the realization of a high sensitivity and a good resolution dictates the choice of the detection method. The most important technique developed for increasing sensitivity and improving resolution is the so-called phase-sensitive detection, associated with an appropriate field modulation. It permits the recording of the derivative of the microwave power absorption with respect to the magnetic field, thereby improving resolution. Resolution is of paramount importance in EPR detection, for much too often the separation between absorption lines is of the order of the line widths, in which case a direct recording of the absorption level hardly discloses any structure in the total line shape. Unfortunately, achievement of the highest sensitivity is invariably accompanied by distortion in the line shape, so that a compromise must be sought according to specific needs.

The effect of modulating the magnetic field may be seen in Fig. 8.5. The magnetic field is moved slowly across the absorption line. At the same time, a small a.c. component applied by an additional coil makes the magnetic field oscillate about the mean instantaneous value. This modulation, while the magnetic field is sweeping the absorption line, results in an oscillation of the absorption of the same frequency (audio or radiofrequency), as shown in Fig. 8.5. Clearly, the oscillation of the absorption is maximum at the inflexion points (a) and vanishes at the top of the absorption curve (b) if the amplitude of the modulation is small enough. The total recording of the process is illustrated by curve (c). There is a phase difference between the oscillating absorption signal before and after the

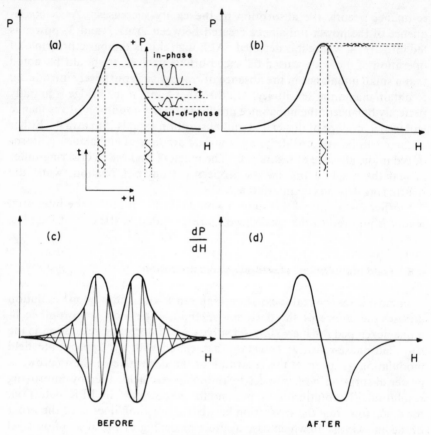

FIG. 8-5 Phase-sensitive detection.

maximum absorption: as long as absorption increases with the magnetic field, the oscillating absorption outphases by π the field modulation (a). If the original modulation signal is multiplied by the oscillating absorption signal, which is precisely the method of phase-sensitive detection, one obtains curve (d), approaching the first derivative of the absorption signal for small amplitudes of the field modulation.

Fig. 8.6 illustrates the incompatibility between high sensitivity and distortionless derivatives. At the top, a small amplitude modulation results in a low-level signal very close to a first derivative. At the bottom, a large amplitude modulation results in a high-level, highly distorted signal. Fig. 10.2 (p.167) shows improvement in resolution achieved by recording the first and second derivatives of an absorption curve consisting of two

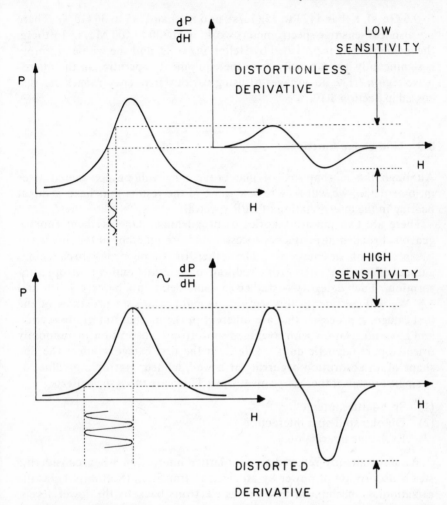

FIG. 8-6 Influence of the modulation amplitude on the shape of the recorded line shape.

partially overlapping absorption lines. Some workers use the technique of recording the second derivative by introducing a second modulation. The improvement in resolution is remarkable but is, of course, achieved at the expense of more elaborate and sometimes highly sophisticated EPR devices.

Commercially available equipments are usually provided with one modulation in the radiofrequency range (e.g., 100 kc/s) and phase-sensitive detection. They work in three microwave frequency bands: X-band (8.5

to 9.6 Gc/s), K-band (23 to 25 Gc/s), and Ka-band (35 to 36 Gc/s). There are also less sensitive spectrometers working at 300 to 400 Mc/s. In these, the magnetic field is provided by Helmholtz coils, and the whole assembly is significantly less expensive than the equipments operating in the microwave region. The advantages of using various frequency ranges are discussed in Section 8.10.

8.9 Line-shape functions

Although the mechanisms of line broadening will be considered later in some detail, we will now review some of the results that have a direct bearing in the interpretation of EPR spectra.

There are two main categories of broadening. One of them (homogeneous broadening) gathers processes that are inherent in the line itself, irrespective of orientation. The other (inhomogeneous broadening) gathers effects that arise from random orientation and anisotropic absorption. Each category is studied in some detail in Chapters 9, 10, and 12. We will consider in this section the main results on processes of the first category: processes that are inherent in the mechanism of absorption and have no relation with averaged anisotropic absorption in randomly oriented paramagnetic units. Effects in the first category affect the line shape of an absorption occurring at a well-defined, isotropic g-value, or in single crystals if there is g-anisotropy. There are three such effects:

(1) Spin-lattice interaction
(2) Dipolar spin-spin interaction
(3) Exchange interaction

We have already referred to spin-lattice interaction when considering steady absorption of power by an electron transition (Section 9.1), as the radiationless mechanism that brings electrons back to the lower levels. Since this process is finite, it contributes to the line broadening and may be assigned a relaxation time τ defined as the time taken for the spin system to lose $(1/e)th$ of its excess energy. The dipolar spin-spin interaction arises from the fact that each electron spin experiences the combined effect of the external magnetic field and the internal magnetic fields created by other spins due to electrons and nuclei (Section 1.6). Exchange interaction occurs when electrons are exchanged between the orbitals of different molecules, and results in narrowing of the absorption line, if the species are alike.

Exchange narrowing leads to Lorentzian line shapes, while the other types of broadening result in Gaussian shapes. Table 8.1 displays the main characteristics of each line shape. Here, T_2 is defined as the inverse

TABLE 8.1. EPR line shapes.

Line shape	Gaussian	Lorentzian
$f(\omega - \omega_0) =$	$\dfrac{T_2}{\pi} \quad \exp\left[-\dfrac{(\omega - \omega_0)T_2^2}{\pi}\right]$	$\dfrac{T_2/\omega}{1 + (\omega - \omega_0)^2 T_2^2}$
width at half height	$(\pi \ln 2)^{1/2} \dfrac{1}{T_2}$	$\dfrac{1}{T_2}$
width at maximum slope	$\dfrac{1}{T_2} (\pi/2)^{1/2}$	$\dfrac{1}{T_2} 3^{-1/2}$

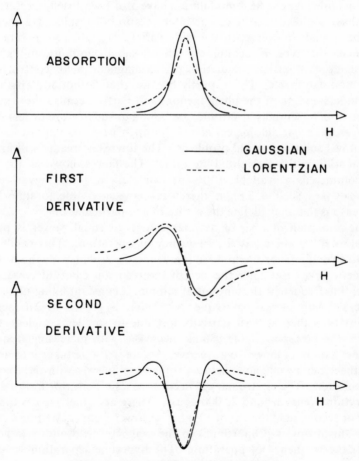

FIG. 8-7 Gaussian and Lorentzian line shapes.

of the line width parameter and is equal to π times the maximum value of $f(\omega - \omega_0)$, i.e.,

$$T_2 = \pi f_{MAX} \tag{8-52}$$

Fig. 8.7 shows typical Gaussian and Lorentzian line shapes together with their derivatives.

8.10   Choice of equipment

So far the essential features of EPR detection have been considered. Some details have been given about resonant cavities, since they are needed for understanding the process of resonance. Phase-sensitive detection and magnetic field modulation have also been briefly explored, for they have a direct bearing on the interpretation of results. No incursion has been made, however, in the wide variety of auxiliary operations inherent in this type of technique; such as, signal amplification, klystron frequency stabilization, regulation of the magnetic field, control of the field modulation, etc. This is partly because they do not add anything to the understanding of the phenomenon, and partly because the details of realization, (essentially electronic) are bound to suffer more or less drastic changes as new technological conquests open new possibilities and provide novel solutions to old problems. The interested reader will find this information in the specialized literature. The fact is, however, that there are commercially available EPR spectrometers working in various frequencies (see Section 8.8), it therefore being advisable to state briefly the reasons that may dictate the choice of any particular one.

The minimum number of detectable spins at equal power is proportional to $\nu^{-7/2}$, where ν is the frequency of operation. This result alone would strongly favor the highest frequency consistent with the available magnetic field. But since the size of the resonant cavity decreases with increasing frequency, the amount of sample, at constant filling factor, also decreases with increasing frequency. What is gained by an increased sensitivity is then at least partially lost due to smaller samples. In addition, the power available usually decreases with increasing frequency. The net gain is no longer so attractive, especially if one takes into account that the available magnetic field *must* increase linearly with the frequency of operation. The electromagnet itself imposes a limit, for ferromagnetic saturation occurs around 20,000 gauss. There are other factors that may limit or recommend the use of a higher or lower frequency. For example, the g-anisotropy is proportional to the magnetic field and is larger, the larger the frequency of operation. The hyperfine separation is, on the other hand, independent of the magnetic field (if this is reasonably large, which is usually the case). When the spectrum has g-anisotropy and hyperfine structure, each particular case will be best handled in a different band. If one is interested in measuring the g-anisotropy, a high frequency would certainly enhance the resolution. If the g-anisotropy is moderate, a low frequency would be more useful for determining the hyperfine interval, since a low frequency tends to minimize the g-anisotropy contribution to the line shape. The type of specimen, even without knowing its EPR spectrum, may influence the choice of the frequency. EPR of conduction electrons in metals may be observed in colloidal metallic particles. However, high frequencies are, in this case, ruled out because their skin depth is much smaller than the size of the colloidal particles, and only frequencies of a few hundred megacycles may be used.* Water has a short dielectric relaxation time; aqueous solutions must therefore be studied at frequencies significantly below 10 Gc/s in order to have a reasonable sensitivity. Many situations may arise and the choice of the best equipment for the application is not always easy, especially taking into consideration that EPR spectrometers are indeed expensive units.

Cited and general references are listed on page 207, at the end of Chapter 12.

---

*At high frequencies, very little electromagnetic field is produced inside a conductor. The process of conduction takes place in a thin layer at the surface whose thickness is called *skin depth*. The skin depth decreases with increasing frequency.

9 The spin Hamiltonian. I. Zeeman term

9.1 The complete Hamiltonian operator

The Hamiltonian operator of a system, it will be recalled, is a quantum mechanical operator whose eigenvalues are the energies of the states of the system. If the system is a paramagnet* in the absence of a magnetic field, the complete Hamiltonian operator is

$$\mathcal{H} = T + V_C + V_{SO} + V_{SS} + V_{SI} + V_X \tag{9-1}$$

where

$$T = \sum_k (p_k^2/2m_0) \tag{9-2}$$

is the total kinetic energy of the electron k of momentum p_k and mass m_0, and the V-terms are potential energy terms, to follow.

---

*The term *paramagnet* has been coined to include paramagnetic ions, atoms, molecules, and molecular fragments in general.

The coulombian term V_C consists of

$$V_C = -\sum_k \frac{Ze^2}{r_k} + \sum \frac{e^2}{r_{ij}} \qquad (9\text{-}3)$$

The first term of Eq. (9-3) is the coulombian attraction between the nucleus and the electrons, the summation being extended over the index k designating each electron of the paramagnet. The second term of Eq. (9-3) represents the coulombian repulsion between electrons and is summed only over pairs of electrons. The third term, V_{SO}, is the potential energy due to the spin-orbit coupling studied in Section 2.3, and is written

$$V_{SO} = \Sigma \lambda_{ij} \mathbf{l}_i \cdot \mathbf{s}_j \qquad (9\text{-}4)$$

where i and j range over all electrons. If one assumes Russell-Saunders coupling, then

$$\Sigma \mathbf{l}_i = \mathbf{L} \qquad (9\text{-}5)$$

and

$$\Sigma \mathbf{s}_j = \mathbf{S} \qquad (9\text{-}6)$$

The vectors **L** and **S** couple to give the total angular momentum

$$\mathbf{J} = \mathbf{L} + \mathbf{S} \qquad (2\text{-}16)$$

and Eq. (9-4) takes the form

$$V_{SO} = \lambda \mathbf{L} \cdot \mathbf{S} \qquad (9\text{-}7)$$

where λ is the spin-orbit coupling constant introduced in Section 2.3.

The fourth term of equation (9-1), V_{SS}, represents the magnetic dipole-dipole interaction between electrons. It is written

$$V_{SS} = \sum_{jk} \frac{1}{r_{jk}^3} \left[\mathbf{s}_j \cdot \mathbf{s}_k - \frac{3\mathbf{s}_j \cdot \mathbf{r}_{jk} \mathbf{r}_{jk} \cdot \mathbf{s}_k}{r_{jk}^2} \right] \qquad (9\text{-}8)$$

The sum, as in the case of coulombian repulsion, is extended over all pairs of electrons. The expression in brackets is obtained by substituting the spin vectors $\mathbf{s}_j$ and $\mathbf{s}_k$ for the magnetic moments $\mathbf{m}_A$ and $\mathbf{m}_B$ in Eq. (1-50) of Section 1.6. The expression $(\mathbf{r}_{jk} \mathbf{r}_{jk})$ without any symbol of multiplication is known as the dyadic form of a second-rank tensor. It operates in the following manner: the scalar product of $\mathbf{s}_k$ by $\mathbf{r}_{jk}$ gives a scalar, which multiplies the numerical value of the vector $\mathbf{r}_{jk}$ without affecting its direction. Thus, the vector **s** is transformed into another vector in the direction of $\mathbf{r}_{jk}$, equal to $\mathbf{r}_{jk} \mathbf{r}_{jk} \cdot \mathbf{s} = (\mathbf{r}_{jk} \cdot \mathbf{s}) \mathbf{r}_{jk}$.

The fifth term of Eq. (9-1) represents the magnetic interaction between unpaired electrons and the nuclear magnetic moments. It consists of two parts: an isotropic part (Fermi or contact interaction) that depends on the electron density at the nucleus under consideration, and an anisotropic part, of the type of interaction studied in Eq. (9-8), but between electrons and nuclei. The Fermi or contact term depends on the *s*-admixture in the state of the electron and is proportional to

$$g g_N \mu_B \mu_N (8\pi/3) \delta(r_k) \mathbf{s}_k \cdot \mathbf{I}_j \tag{9-9}$$

where g_N and μ_N are the nuclear *g*-factor and magneton, $\mathbf{I}_j$ is the nuclear spin vector operator, $\mathbf{r}_k$ is the distance of electron k to nucleus j, and $\delta(r_k)$ is the Dirac function. The anisotropic term, which depends on the amount of admixture other than *s* in the state of the electron, is expressed by

$$g g_N \mu_B \mu_N \left[\frac{\mathbf{s}_k \cdot \mathbf{I}_j}{\mathbf{r}_k^3} - \frac{3 \mathbf{s}_k \cdot \mathbf{r}_k \mathbf{r}_k \cdot \mathbf{I}_j}{\mathbf{r}_k^5} \right] \tag{9-10}$$

Clearly, the V_{SI} term (also known as the hyperfine term) is obtained by the summation of Eqs. (9-9) and (9-10) over all the unpaired electrons k and nuclei j. (This is studied in Chapter 10.)

If the nucleus has a spin I greater than one-half, quadrupole interactions take place. These effects will be disregarded for the present.

Finally, the last term, V_X, represents the interaction of the paramagnet with the crystal field, being essentially electrostatic. It is usually written

$$V_X = -\sum_k e_k \Phi(\mathbf{r}_k) \tag{9-11}$$

where the summation is extended over all electrons. The study of the V_X term is the subject of Chapter 12.

The order of magnitude of these terms for a typical paramagnetic ion of the iron group, for example, is, in wavenumbers expressed in cm^{-1}: $V_C \approx 10^5$, $V_{SO} \approx 10^2$, $V_{SS} \approx 10^0$, $V_{SI} \approx 10^{-2}$, $V_X \approx 10^3$ or greater if there is considerable covalent character in the bonds between the central ion and the ligand molecules. From these figures, one finds that the energies associated with V_C lie in the far ultraviolet, and those of V_X may fall in the visible and near ultraviolet, thus accounting for optical transitions characterizing the color of crystals and solutions of most paramagnetic ions.

9.2 The spin Hamiltonian

The Hamiltonian, in the general form established in Section 9.1, is exceedingly difficult to work with. It is possible, however, to arrive at a much simpler expression by applying perturbation theory. The derivation is out of the scope of this book and may be found in more specialized publications.[5] The result is the so-called "effective spin Hamiltonian," or spin Hamiltonian for brevity:

$$\mathcal{H} = \mu_B \mathbf{S} \cdot \mathbf{g} \cdot \mathbf{H} + \mathbf{S} \cdot \mathbf{A} \cdot \mathbf{I} + \mathbf{S} \cdot \Phi \cdot \mathbf{S} - g_N \mu_N \mathbf{I} \cdot \mathbf{H} \qquad (9\text{-}15)$$

where **g**, **A**, and Φ are second-rank tensors, and **S** is now the *effective electron spin operator,* whose value is determined by putting the multiplicity of the ground state equal $(2S + 1)$. The first term of Eq. (9-15) is called Zeeman term and takes the place of Eq. (9-4) and (9-12). The **g**-tensor, which now takes the place of the usual spectroscopic splitting factor, is a symmetric tensor whose elements are composed of the free-electron g-value plus an anisotropic correction due to the spin-orbit coupling. The **A**-tensor describes the magnetic hyperfine interaction and is, although only partially, equivalent to the combined effects represented by Eqs. (9-9) and (9-10), since the spin operator appearing in these equations is the actual spin of the electron, while the spin Hamiltonian refers to the effective electronic spin of the paramagnet. The **A**-tensor contains an isotropic contribution arising from the Fermi contact term (Eq. 9-9) and an anisotropic contribution due to the spin-orbit coupling and a pure orbital effect. In most cases, the symmetry of both **g**- and **A**-tensors is intimately related to the symmetry of the environment. When both **g** and **A** have axial symmetry, they are diagonalized by the same system of coordinates; otherwise this may not be the case, a situation that considerably complicates the interpretation of results. The Φ-tensor represents the removal of spin degeneracy for $S \geq 1$ (zero-field splitting) by a noncubic crystal field. It again contains a contribution from the spin-orbit coupling and a usually smaller contribution from the spin-spin interaction. Each of these terms is studied in following sections.

9.3 The Zeeman term: Anisotropic case

If zero-field splittings and nuclear effects are ignored, the spin Hamiltonian reduces to

$$\mathcal{H} = \mu_B \mathbf{S} \cdot \mathbf{g} \cdot \mathbf{H} \qquad (9\text{-}16)$$

where **g** is a symmetric tensor

$$\mathbf{g} = \begin{bmatrix} g_{11} & g_{12} & g_{13} \\ g_{12} & g_{22} & g_{23} \\ g_{13} & g_{23} & g_{33} \end{bmatrix} \qquad (9\text{-}17)$$

There is a reference frame that diagonalizes **g**. Eq. (9-17) then becomes

$$\mathcal{H} = \mu_B [S_x \ S_y \ S_z] \cdot \begin{bmatrix} g_1 & 0 & 0 \\ 0 & g_2 & 0 \\ 0 & 0 & g_3 \end{bmatrix} \cdot \begin{bmatrix} H_x \\ H_y \\ H_z \end{bmatrix}$$

$$\mathcal{H} = \mu_B (g_1 S_x H_x + g_2 S_y H_y + g_3 S_z H_z) \tag{9-18}$$

Expression (9-18) may be interpreted as follows: the **g**-tensor operates on the **H** column vector to give an effective field at the electron which is, in column vector form:

$$\begin{bmatrix} g_1 H_x \\ g_2 H_y \\ g_3 H_z \end{bmatrix} \tag{9-19}$$

Scalar multiplication of this column vector by the spin operator row vector

$$\mathbf{S} = [S_x \ S_y \ S_z] \tag{9-20}$$

gives the scalar product of vectors expressed in Eq. (9-18) if performed *from right to left*,* i.e.,

$$\mathcal{H} = \mu_B (g_1 S_x H_x + g_2 S_y H_y + g_3 S_z H_z) \tag{9-21}$$

An alternate interpretation is to consider the row vector $\mu_B \mathbf{S} \cdot \mathbf{g}$ as the effective magnetic moment of the electron, in which case Eq. (9-21) is also arrived at.

The energy of the levels will then be given by

$$E = \mu_B S_H H (g_1^2 \sin^2 \theta \cos^2 \phi + g_2^2 \sin^2 \theta \sin^2 \phi + g_3^2 \cos^2 \theta)^{1/2} = \mu_B g_H S_H H \tag{9-32}$$

with the meaning of the angles indicated in Fig. 9.1. Notice the analogy between this treatment and the treatment of paramagnetic anisotropy developed in Chapter 7. The reason is clear: in thermal equilibrium, it is the difference in population of the energy levels of Eq. (9-32) that produces the magnetization studied in Chapter 7. Therefore, the *g*-value in the direction of **H**

$$g_H = (g_1^2 \sin^2 \theta \cos^2 \phi + g_2^2 \sin^2 \theta \sin^2 \phi + g_3^2 \cos^2 \theta)^{1/2} \tag{9-33}$$

determines the value of the susceptibility along the same direction:

$$\kappa_H = \kappa_1 l^2 + \kappa_2 m^2 + k_3 n^2 \tag{7-37}$$

where l, m, and n are the direction cosines introduced in Eq. (7-27).

---

*The product of a column vector by a row vector is not commutative. Multiplication from right to left of a column vector by a row vector gives a scalar; multiplication from right to left of a row vector by a column vector gives a square matrix.

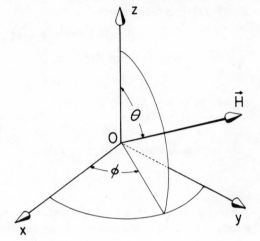

FIG. 9-1 Polar and azimuthal angles.

If the **g**-anisotropy is not exceedingly large, i.e., if

$$|g_2 - g_1| \ll g_1, \qquad |g_3 - g_1| \ll g_1 \qquad (9\text{-}34)$$

Eq. (9-33) may be written

$$g_H = g_1 + (g_2 - g_1)\sin^2\theta\cos^2\phi + (g_3 - g_1)\cos^2\theta \qquad (9\text{-}34)$$

neglecting higher-order terms in $\Delta g/g$. As in the case of κ, an anisotropic **g**-tensor is usually associated with symmetry environments that do not exchange x, y, or z (orthorhombic and less symmetric fields), although (also as in the case of κ) the anisotropy of **g** may be due to the Jahn-Teller effect (Section 2.7).

In the liquid phase, the paramagnets are tumbling and rotating; the different g-values average to

$$\langle g^2 \rangle = \frac{1}{3}\,(g_1^2 + g_2^2 + g_3^2) \qquad (9\text{-}36)$$

The case of polycrystalline samples is considered in Section 9.5.

9.4   The Zeeman term: Axial and isotropic cases

In a symmetric environment in which there are operations of symmetry that exchange x and y but not z (tetragonal, rhombohedral, and hexagonal fields), one can show (as in the case of the κ-tensor) that the **g**-tensor has axial symmetry in the absence of Jahn-Teller distortion; i.e.,

$$g_1 = g_2 = g_\perp \qquad (9\text{-}37)$$

$$g_3 = g_\parallel \qquad (9\text{-}40)$$

Eq. (9-33) then reduces to

$$g_H = (g_\perp^2 \sin^2 \theta + g_\parallel^2 \cos^2 \theta)^{1/2} \qquad (9\text{-}39)$$

If the anisotropy is not large, i.e., if

$$|g_\perp - g_\parallel| \ll g_\perp, g_\parallel \qquad (9\text{-}40)$$

Eq. (9-39) may be further simplified to

$$g_H = g_\perp + (g_\parallel - g_\perp) \cos^2 \theta \qquad (9\text{-}41)$$

In the liquid state, the average value takes the form

$$\langle g^2 \rangle = \frac{1}{3} (2g_\perp^2 + g_\parallel^2) \qquad (9\text{-}42)$$

Finally, in a cubic environment with no Jahn-Teller distortion, the **g**-tensor reduces to the scalar g-value and the Zeeman term becomes

$$\mu_B g \mathbf{S} \cdot \mathbf{H} = \mu_B g\, S_H H \qquad (9\text{-}43)$$

Clearly, Eq. (9-43) holds for s-electrons, and for electrons with the orbital momentum completely quenched.

9.5 Experimental determination of the principal g-values

The determination of the principal g-values is not as simple as the determination of the magnetic axes due to two reasons. On one hand, not always is the g-anisotropy trivially related to the symmetry of the crystal. On the other hand, coexistence of differently oriented paramagnets often makes the problem particularly difficult. In general, the procedure consists of finding the directions for which the g-value has the maximum and the minimum values. The third value is easily found in a direction perpendicular to the plane that contains the maximum and minimum values. Once the orientations are known, the problem is reduced to measuring the energy of the transitions $\Delta m_S = 1$ between the Zeeman energy levels by aligning the magnetic field with each canonical orientation in turn. The eigenvalues of the spin Hamiltonian of Eq. (9-16) then become

$$E_1 = \mu_B g_1 m_S H_x \quad \text{for} \quad H_x = H \qquad (9\text{-}44)$$

$$E_2 = \mu_B g_2 m_S H_y \quad \text{for} \quad H_y = H \qquad (9\text{-}45)$$

$$E_3 = \mu_B g_3 m_S H_z \quad \text{for} \quad H_z = H \qquad (9\text{-}46)$$

for each value of $m_S = S_H$. This procedure is simple if one deals with a paramagnet uniformly oriented in a single crystal. If this is not the case, a comparative study of the change with orientation of the line shapes in a single crystal may lead to the principal values, although in this case some experience in interpretation of spectra is needed. Curiously enough, in the case of moderately large **g**-anisotropy a study of polycrystalline material leads to those values.[10] In order to understand the basic principle behind this method, it is necessary to analyze the situation that takes place in polycrystalline material. First consider an axially symmetric **g**-tensor. The z-axes of the paramagnets point at random. However, as a consequence of random orientation, there are more paramagnets with their z-axes close to perpendicular to the direction of the magnetic field (intersecting the area $2\pi R^2 d\theta$ in Fig. 9.2) than there are parallel (lying in the solid angle $d\Omega$).[11] The density of population at orientation θ is propor-

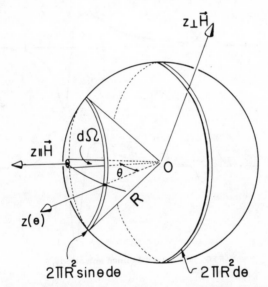

FIG. 9-2   Orientations of an axially symmetric paramagnet.

tional to the distribution function $\sin\theta$. Fig. 9.3 illustrates this state of affairs in the case of $Ag(o\text{-phn})_2S_2O_8(^2D_{21/2})$. At the top (a) the theoretical population distribution is shown in dotted lines, as well as the power absorption curve, smoothed by broadening processes. Curves (b) and (c) are the first and second derivatives of the power with respect to the magnetic field.[13]

The case of total anisotropy is even more interesting. Since the value of g_2 lies between those of g_1 and g_3, there are orientations at which g_H

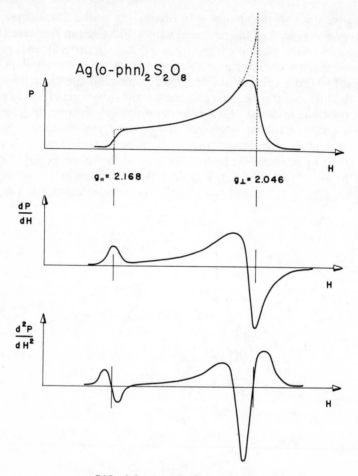

FIG. 9-3   Axial Zeeman anisotropy.

is numerically equal to g_2, though having g_1 and g_3 contributions. In particular, there is an orientation in the plane xz at which

$$g = g_2 = (g_1^2 \sin^2 \theta_0 + g_3^2 \cos^2 \theta_0)^{1/2} \tag{9-46}$$

without any contribution from g_2. All the orientations lying in the plane that contains $g(\theta_0, \phi = 0)$ and g_2 have the same g-value, as illustrated in Fig. 9.4. The value of θ_0 may be derived as follows.

By introducing the expressions

$$g_1 = g_2 - \delta_1, \qquad g_3 = g_2 + \delta_2 \tag{9-47}$$

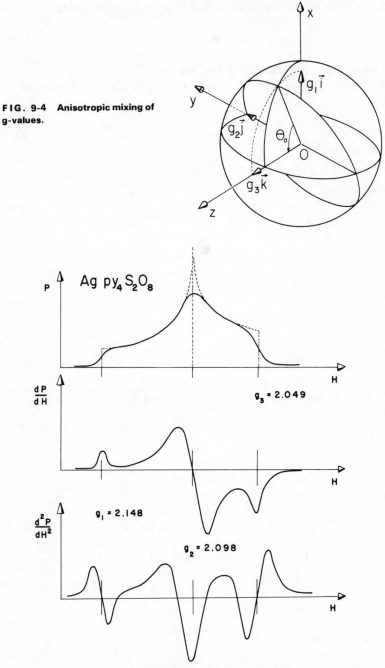

FIG. 9-4   Anisotropic mixing of g-values.

FIG. 9-5   Total Zeeman anisotropy.

in Eq. (9-46), neglecting terms of higher order in δ, one obtains

$$\theta_0 = \tan^{-1}\left[\frac{g_2 - g_1}{g_3 - g_2}\right]^{1/2} \qquad (9\text{-}48)$$

A typical case[13] is illustrated in Fig. 9.5.

Cited and general references are listed on page 207, at the end of Chapter 12.

10 *The spin Hamiltonian.*
II. Hyperfine term

10.1 Electron spin-nuclear spin interaction

The existence of nuclei with magnetic moments other than zero (nuclear spin $I = \frac{1}{2}, 1, 1\frac{1}{2}, \ldots$) produces a splitting of the Zeeman levels given by Eq. (9-16). This is due to the fact that the nuclear spins also quantize in the magnetic field at the nucleus. Thus, the electron experiences the combined effect of the external magnetic field and the field at the electron due to the nuclear magnetic moments, in addition to a variable direct coupling when the electronic density at the nucleus is not zero (*s*-admixture). This effect is represented by the term

$$\mathbf{S} \cdot \mathbf{A} \cdot \mathbf{I} \qquad\qquad 10\text{-}1)$$

where, as before, $\mathbf{S}$ and $\mathbf{I}$ are the effective electron spin and the nuclear spin vector operators, and $\mathbf{A}$ is a symmetric, second-rank tensor called hyperfine coupling tensor. Since the difference between the nuclear magnetic energy levels is, under prevailing experimental conditions, about three orders of magnitude smaller than the difference between the electron levels, the population of the nuclear levels is almost equal. By the same reasoning, the resonance frequency of the electronic transitions ($\Delta m_S = \pm 1$) cannot affect the population of the nuclear levels and the EPR permitted transitions occur at $\Delta m_I = 0$. Each electron Zeeman level given by the term

$$\mu_B \mathbf{S} \cdot \mathbf{g} \cdot \mathbf{H} \tag{9-16}$$

is then split into $2I + 1$ levels, corresponding to $m_I = -I, -(I - 1), \ldots, +I$, which determine an equal number of electron transitions at $\Delta m_S = 1$ and $\Delta m_I = 0$. If the Zeeman term is large as compared with the hyperfine term (i.e., if the external magnetic field is strong), the coupling tensor **A** does not depend on **H**. However, for small magnetic fields, the Zeeman and hyperfine terms become the same order in magnitude, and expression (10-1) can no longer be applied due to the now comparatively strong **I-S** contact interaction coupling resulting in an effective spin

$$\mathbf{F} = \mathbf{S} + \mathbf{I} \tag{10-2}$$

which no longer identifies with the effective electron spin used thus far. This state of affairs is illustrated in Fig. 10.1 for $S = I = \frac{1}{2}$ and was

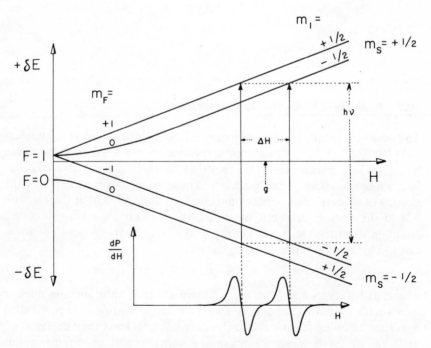

FIG. 10-1 Electron spin-nuclear spin zero-field coupling and hyperfine splitting.

quantitatively studied by Breit and Rabi.[7] At the left it is seen that the levels do not converge to a single value but approach the zero-field energy levels for $F = 1$ and $F = 0$. In this region, transitions occur at $\Delta m_F = \pm 1$ if the oscillating magnetic field H_1 is perpendicular to **H** and at $\Delta m_F = 0$ if parallel. Additional information may then be gained if it is possible to operate the resonance equipment at rather low frequencies.[21] However, the bulk of EPR research is done under conditions of "strong field," which means that Eq. (10-1) can be applied almost without restrictions.

As in the case of the Zeeman tensor, the hyperfine tensor may be isotropic, in which case it reduces to the scalar A according to

$$A \mathbf{S} \cdot \mathbf{I} \tag{10-3}$$

The anisotropic cases are studied in the following sections.

10.2   Anisotropic and axial hyperfine term

If the hyperfine coupling is totally anisotropic, there will be, as in the case of the **g**-tensor, a reference frame that diagonalizes its matrix representation. In this frame, Eq. (10-1) takes the form

$$\mathcal{H}_{hf} = [S_x \; S_y \; S_z] \cdot \begin{bmatrix} A_1 & 0 & 0 \\ 0 & A_2 & 0 \\ 0 & 0 & A_3 \end{bmatrix} \cdot \begin{bmatrix} I_x \\ I_y \\ I_z \end{bmatrix} \tag{10-4}$$

and the hyperfine coupling energies along the principal hyperfine axes are the eigenvalues

$$E_{1(hf)} = A_1 m_S m_I \tag{10-5}$$

$$E_{2(hf)} = A_2 m_S m_I \tag{10-6}$$

$$E_{3(hf)} = A_3 m_S m_I \tag{10-7}$$

at the canonical orientations $H = H_x$, $H = H_y$, and $H = H_z$. Unfortunately, hyperfine anisotropy is usually accompanied by Zeeman anisotropy. In such a case, the same reference frame does not necessarily diagonalize all tensors, and a rather difficult analysis of the dependance of resonance with orientation is needed.[12,16] This may be particularly cumbersome if, in addition, there are paramagnets oriented in different ways, a common situation found in irradiation-produced radicals.

For axial symmetry, Eq. (10-4) reduces to

$$\mathcal{H}_{hf} = [S_x \; S_y \; S_z] \cdot \begin{bmatrix} A_\perp & 0 & 0 \\ 0 & A_\perp & 0 \\ 0 & 0 & A_\parallel \end{bmatrix} \cdot \begin{bmatrix} I_x \\ I_y \\ I_z \end{bmatrix} \tag{10-8}$$

if the reference frame diagonalizes **A**. Fortunately enough, axial hyperfine symmetry is accompanied by axial Zeeman symmetry and the same reference frame diagonalizes both tensors. The effective spin Hamiltonian for $S = \frac{1}{2}$ may then be written, neglecting nuclear Zeeman and spin-spin interactions, as

$$\mathcal{H} = \mu_B[g_\perp^2 \sin^2\theta + g_\parallel^2 \cos^2\theta]^{1/2}\,\mathbf{H}\cdot\mathbf{S} + A_\parallel S_z I_z + A_\perp[S_x I_x + S_y I_y]$$

$$= \mu_B g_H \mathbf{H}\cdot\mathbf{S} + K\mathbf{S}\cdot\mathbf{I} \tag{10-9}$$

where g_H is given by Eq. (9-39) and K is

$$K = \left[\frac{A_\parallel^2 g_\parallel^2}{g_H^2}\cos^2\theta + \frac{A_\perp^2 g_\perp^2}{g_H^2}\sin^2\theta\right]^{1/2} \tag{10-10}$$

In the canonical orientations, the eigenvalues of the spin Hamiltonian are clearly given by

$$E_\perp = \mu_B g_\perp H m_S + A_\perp m_S m_I \tag{10-11}$$

$$E_\parallel = \mu_B g_\parallel H m_S + A_\parallel m_S m_I \tag{10-12}$$

for each pair of values (m_S, m_I).

Eq. (10-9) is an approximation that leads to the equally spaced energy levels of Eqs. (10-11) and (10-12). In higher approximations, the lines are not exactly equidistant and K is a function of θ, and of θ and ϕ in the more general case of total anisotropy. Interested readers will find a detailed account of this state of affairs in Low's book.

The determination of the hyperfine tensor principal values must be done by studying single crystals. If the **g**-tensor is axial and the g-anisotropy is significantly larger than the hyperfine interval (strong magnetic field), the hyperfine tensor principal values may be evaluated in samples having randomly oriented paramagnets, especially if they are magnetically dilute so that electronic spin-spin interactions are negligible. This is the case of some frozen solutions. Fig. 10.2 shows the EPR spectrum of $Ag^{2+}(^2D_{21/2})$ in frozen acid solution.[13] Silver has two isotopes, both of spin $I = \frac{1}{2}$, differing in nuclear magnetic moment by less than 10 percent, which simplifies the problem and permits its treatment, in first approximation, as an unpaired electron interacting with a nucleus of spin one half.

Strong hyperfine interaction in randomly oriented paramagnetic species gives rise, if the g-anisotropy is small, to situations similar to those involving strong g-anisotropy in the absence of hyperfine interaction. They need not be discussed here, for they lead to absorption curves closely resembling those due to g-anisotropy alone.

The existence of a strong, total Zeeman anisotropy presents a some-

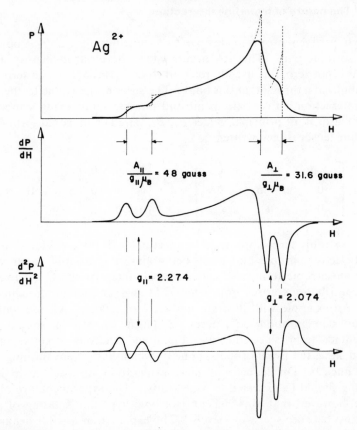

FIG. 10-2 Hyperfine structure disclosed by strong Zeeman aniso-tropy.

what different problem. If the Zeeman and hyperfine tensors are di-agonalized by the same choice of coordinates, the extremes of the spec-trum, at g_1 and g_3, will display the hyperfine principal values A_1 and A_3. The center g-value will, however, show the combined effect of the canoni-cal splitting A_2 and the various A_1 and A_3 contributions arising from the orientations that lie in the plane referred to in Eq. (9-46). When the hyper-fine interaction is mainly due to the contact term (see next section), it is possible to solve completely the problem. If different coordinate systems are needed to diagonalize each tensor, the hyperfine splitting at the ends of the spectrum will have off-diagonal terms and the problem cannot be solved without additional information.

10.3   The nature of hyperfine interaction

The hyperfine tensor is made up of two contributions: an isotropic and an anisotropic part. The first occurs when the electron density at the nucleus is not zero; it is usually referred to as Fermi or contact term, and is symbolized a throughout this book. The second one is due to the magnetic interaction between electronic and nuclear spins, being symbolized by the tensor $\mathbf{B}$, of principal values B_1, B_2, B_3. The diagonal form of the hyperfine tensor is then written

$$
\mathbf{A} = \begin{bmatrix} a + B_1 & 0 & 0 \\ 0 & a + B_2 & 0 \\ 0 & 0 & a + B_3 \end{bmatrix} = a + \begin{bmatrix} B_1 & 0 & 0 \\ 0 & B_2 & 0 \\ 0 & 0 & B_3 \end{bmatrix} = a + \mathbf{B} \quad (10\text{-}13)
$$

The isotropic term appears in connection with the s-character of unpaired electrons in ions and atoms being almost invariably accompanied by the anisotropic term, also called dipolar interaction. Coexistence of both contributions arises from what is known as configuration interaction. An example will illustrate this state of affairs. An ion with the electron configuration $3d^5$, term ^6S (such as Mn^{2+}) should have no contact term since the five unpaired electrons occupy, in the ground state, the five d-orbitals, none of which has a nonvanishing value at the nucleus. On the basis of this consideration, no isotropic hyperfine splitting should be expected in Mn^{2+} salts.[1] The experimental results indicate, however, that Mn^{2+} has an isotropic hyperfine splitting of about 100 gauss between successive levels.[4] The explanation for this behavior is that although the ground configuration of Mn^{2+} is $3s^2\text{-}3p^6\text{-}3d^5$, there is an excited state $3s^1\text{-}3p^6\text{-}3d^5\text{-}4s^1$ with two unpaired s electrons. Interaction between the two configurations naturally leads to the existence of contact term, arising from the s admixture due to the excited state. Interaction of an s electron with the nucleus is very large (usually of a thousand gauss or more), so that, in most cases, only a small amount of the excited state need be admixed to account for the observed isotropic splitting. A more rigorous study of d-orbitals shows that the isotropic interaction need not be necessarily due to s-admixture but to core polarization which determines non vanishing wave functions at the nucleus. Thus, ions with unpaired d-orbital electrons exhibit isotropic contributions to the hyperfine interaction.

The quantum mechanical expression for the contact term is[14]

$$
a = a_s^2 g \mu_B g_N \mu_N (8\pi/3) \mathbf{S} \cdot \mathbf{I} \delta(\mathbf{r}_e - \mathbf{r}_N) \quad (10\text{-}14)
$$

already advanced in Eq. (9-9). The coefficient a_s^2 is the probability that the electron shall be in an s state. The other symbols have already been introduced. The Dirac delta function $\delta\,(\mathbf{r}_e - \mathbf{r}_N)$ is expressed here as a function of the vector position of the electron, $\mathbf{r}_e$, and nucleus, $\mathbf{r}_N$. Clearly, the delta function is normalized in the three dimensions. Eq. (10-14) is often written as a function of the value at the nucleus of the normalized electron spin density $\vartheta\,(\mathbf{r}_N)$, i.e.,

$$a = a_s^2 g\mu_B g_N\mu_N(8\pi/3)\vartheta\,(\mathbf{r}_N) \qquad (10\text{-}15)$$

since the delta function is zero for $\mathbf{r}_e \neq \mathbf{r}_N$.

The anisotropic hyperfine term is then given by

$$\mathbf{S}\cdot\mathbf{B}\cdot\mathbf{I} = -(1 - a_s^2)g\mu_B g_N\mu_N \left[\frac{\mathbf{S}\cdot\mathbf{I}}{|\,r\,|^3} - \frac{3\mathbf{S}\cdot\mathbf{rr}\cdot\mathbf{I}}{|\,r\,|^5}\right] \qquad (10\text{-}16)$$

where $(1 - a_s^2)$ stands for the probability that the electron shall not be in an s state. Since the electron spin is described by the density distribution function $\vartheta\,(\mathbf{r}) = \psi^*(\mathbf{r})\psi(\mathbf{r})$ (where $\mathbf{r}$ is now the vector $(\mathbf{r}_e - \mathbf{r}_N)$) the principal values (at the canonical orientations) of the anisotropic contribution to the eigenvalues of the hyperfine Hamiltonian are

$$B_1 = \Omega \int_\epsilon \vartheta\,(\mathbf{r})\,\frac{(3\sin^2\theta'\cos^2\phi' - 1)}{|\,r\,|^3}\,dV \qquad (10\text{-}17)$$

$$B_2 = \Omega \int_\epsilon \vartheta\,(\mathbf{r})\,\frac{(3\sin^2\theta'\sin^2\phi' - 1)}{|\,r\,|^3}\,dV \qquad (10\text{-}18)$$

$$B_3 = \Omega \int_\epsilon \vartheta\,(\mathbf{r})\,\frac{(3\cos^2\theta' - 1)}{|\,r\,|^3}\,dV \qquad (10\text{-}19)$$

with

$$\Omega = (1 - a_s^2)g_H\mu_B g_N\mu_N \qquad (10\text{-}20)$$

In Eqs. (10-17) to (10-19), θ' and ϕ' are the polar and azimuthal angles of the $\mathbf{r}$ vector in the system of coordinates that diagonalizes $\mathbf{B}$, and g_H is the electron g-value in the direction of the magnetic field. The symbol ϵ indicates that a small region about the nucleus is omitted from the integration to avoid divergence. (The expression for g_H is given in Eq. (9-33).) It should be pointed out, however, that in this case, the expression is referred to the system of coordinates that diagonalizes the g-tensor, which is not necessarily the one that diagonalizes the hyperfine tensor. If the same reference frame diagonalizes both tensors, the value of g_H in Eqs. (10-17) to (10-19) must be substituted for by the principal values g_1, g_2, and g_3.

10.4 Hyperfine splitting in liquids

Paramagnets in a liquid phase are rotating and tumbling at a rate related to the viscosity of the medium. If the tumbling frequency and the angular velocity of rotation are higher than the frequency of the oscillating magnetic field (microwave frequency) inside the cavity, the anisotropic hyperfine interaction averages out, leaving only the Fermi interaction. The physical interpretation of Eqs. (10-17) to (10-19) permits one to understand such a behavior, as follows.

Eqs. (10-17) to (10-19) represent the magnetic energy arising from electron dipole-nuclear dipole interaction. They may be put as the scalar product of the effective electron spin magnetic vector by the magnetic field experienced by the electron due to the nucleus, given in Eq. (3-19). The effective electron spin magnetic moment is clearly

$$- (1 - a_s^2) g_H \mu_B m_S \tag{10-21}$$

where we have included the probability factor $(1 - a_s^2)$ to account for the fact that the electron has an s admixture of probability a_s^2, which must be subtracted when dipolar interaction with the central nucleus is under consideration.* The effective magnetic field at the electron due to the nucleus then becomes, in each canonical orientation:

$$h_1 = - m_1 g_N \mu_N \int_\epsilon \vartheta(\mathbf{r}) \frac{(3 \sin^2 \theta' \cos^2 \phi' - 1)}{|r|^3} dV \tag{10-22}$$

$$h_2 = - m_1 g_N \mu_N \int_\epsilon \vartheta(\mathbf{r}) \frac{(3 \sin^2 \theta' \sin^2 \phi' - 1)}{|r|^3} dV \tag{10-23}$$

$$h_3 = - m_1 g_N \mu_N \int_\epsilon \vartheta(\mathbf{r}) \frac{(3 \cos^2 \theta' - 1)}{|r|^3} dV \tag{10-24}$$

The quantities h_1, h_2, and h_3 are the components of the actual magnetic field due to the nuclear magnetic dipole along the direction of the electron spin quantization. These components may be written

$$h_1 = - m_1 g_N \mu_N \int \vartheta(\mathbf{r}) f_1(\mathbf{r}) dV \tag{10-25}$$

$$h_2 = - m_1 g_N \mu_N \int \vartheta(\mathbf{r}) f_2(\mathbf{r}) dV \tag{10-26}$$

$$h_3 = - m_1 g_N \mu_N \int \vartheta(\mathbf{r}) f_3(\mathbf{r}) dV \tag{10-27}$$

*Dipolar interaction with the central nucleus cancels out in an s state due to spherical symmetry.

where

$$f_1(\mathbf{r}) = \frac{(3\sin^2\theta'\cos^2\phi' - 1)}{|r|^3} \tag{10-28}$$

$$f_2(\mathbf{r}) = \frac{(3\sin^2\theta'\sin^2\phi' - 1)}{|r|^3} \tag{10-29}$$

$$f_3(\mathbf{r}) = \frac{(3\cos^2\theta' - 1)}{|r|^3} \tag{10-30}$$

The sum

$$h_1 + h_2 + h_3 = -m_1 g_N \mu_N \int \vartheta(\mathbf{r})[f_1(\mathbf{r}) + f_2(\mathbf{r}) + f_3(\mathbf{r})]\,dv \tag{10-31}$$

is equal to zero since

$$f_1(\mathbf{r}) + f_2(\mathbf{r}) + f_3(\mathbf{r}) = 0 \tag{10-32}$$

Therefore,

$$h_1 + h_2 + h_3 = 0 \tag{10-33}$$

If the paramagnets change their orientation much faster than the value of the oscillating magnetic field inside the cavity, the magnetic field at the electron due to the nuclear magnetic dipole will average out and the sole hyperfine effect left will be the isotropic, Fermi term. This usually happens in the case of liquids. In a sample in which the paramagnets are not changing in orientation, such as a polycrystalline solid or a glass, as long as the orientation of the paramagnets is random, the dipolar term just broadens the absorption lines. This is developed in Chapter 11.

10.5   Interpretation of the hyperfine structure

Eq. (10-33) reduces, in axial symmetry, to

$$2h_\perp + h_\parallel = 0 \tag{10-34}$$

Clearly, no isotropic effect may arise from electron dipole-nuclear dipole interaction, for in such a case

$$h_1 = h_2 = h_3 \tag{10-35}$$

which can only be true, in view of Eq. (10-33), if

$$h_1 = h_2 = h_3 = 0 \tag{10-36}$$

Eqs. (10-33) and (10-34) are useful in the interpretation of spectra. In transition ions (especially ions of the iron group), where the unpaired electrons are not shielded against the forces of the ligand field by outer electrons, and in radicals, the orbital quenching is large and the g-tensor only slightly anisotropic. In these cases the principal g-values approach the free-spin value.

The property that the anisotropic splitting averages out in rapidly rotating molecules is mathematically contained in the trace of the B-tensor, which is zero, i.e.:

$$B_1 + B_2 + B_3 = 0 \tag{10-37}$$

and, in axial symmetry

$$B_\parallel + 2B_\perp = 0 \tag{10-38}$$

Since the hyperfine-tensor principal values are

$$A_1 = a + B_1 \tag{10-39}$$

$$A_2 = a + B_2 \tag{10-40}$$

$$A_3 = a + B_3 \tag{10-41}$$

the contact term a is given by

$$a = \frac{1}{3}(A_1 + A_2 + A_3) \tag{10-42}$$

and in axial symmetry by

$$a = \frac{1}{3}(A_\parallel + 2A_\perp) \tag{10-43}$$

However the determination of a is not always unambiguous unless the relative signs of A_1, A_2 and A_3 are known. Otherwise there are four possible solutions, namely:

$$a = \frac{1}{3}(A_1 + A_2 + A_3)$$

$$a = \frac{1}{3}(-A_1 + A_2 + A_3)$$

$$a = \frac{1}{3}(A_1 - A_2 + A_3) \tag{10-44}$$

$$a = \frac{1}{3}(A_1 + A_2 - A_3)$$

which clearly lead to different values of a and B_i. If the hyperfine splitting is large, additional studies of the intensity lead to sign assignment by considering second-order effects. If the hyperfine splitting is, on the other hand, small, sometimes it is possible to determine relative signs by performing ENDOR and observing the relative intensity of the two ENDOR lines in fast passages at increasing and decreasing radiofrequency. Finally, in some cases it is possible to assume signs on the basis of physical arguments. For example, if the spectrum is only slightly anisotropic one may infer that the sign of A_1, A_2 and A_3 is that of the contact term due to a large s-admixture or high core polarization. This situation is usually encountered in free radicals. If the **A**-tensor is axially symmetric

$$A_\perp - A_\parallel = B_\perp - B_\parallel \qquad (10\text{-}45)$$

and the problem is solved after considering Eq. (10-38). If the **A**-tensor is not axial, the principal values of the **B**-tensor result from Eqs. (10-39), (10-40) and (10-41).

Two are the approximations implied in the derivation of the hyperfine equations, the electron effective spin is quantized in the external magnetic field and the nuclear spin is quantized in the field due to the electron. Such assumptions are justified if the value of the external magnetic field lies between the field at the electron due to the nucleus and the field at the nucleus due to the electron. The assumption that the nuclear spin is quantized in the field due to the electron makes the hyperfine interaction vanish for $m_I = 0$, since in this case the spin vector I is perpendicular to the lines of force of the electron magnetic field. On the other hand, the electron actually quantizes in the magnetic field resulting from the external and the nuclear field. If this latter is much smaller than the external magnetic field, it does not significantly change the direction of quantization of the electron spin, and the electron spin only "sees" the component of the nuclear magnetic field in the direction of the external magnetic field. In turn, each assumption would be invalidated by very large or very small external magnetic fields. Within this approximation, interaction of an electron with a nucleus of spin I gives rise to $2I + 1$ lines arising from the $2I + 1$ possible values of m_I and the selection rule that allowed transitions are those occurring with $\Delta m_S = 1$ and $\Delta m_I = 0$. Since the nuclear levels are almost equally populated, the lines arising from interaction with *one* nucleus all have the same intensity. The lines are, in addition, equidistant. In higher approximations, it is found that the lines are not exactly equidistant, particularly if the nuclear quadrupolar effect is taken into account. If the hyperfine interaction is very small, it may be necessary to take into account the direct interaction of the nuclear spin and the external field, in which case very weak extra lines may appear in the spectrum.

Simultaneous interaction of the unpaired electrons with more than one nucleus leads to a different pattern of lines. The hyperfine Hamiltonian must be put in the form

$$\mathcal{H}_{hf} = \sum_k \mathbf{S} \cdot \mathbf{A}_k \cdot \mathbf{I}_k \tag{10-46}$$

and the summation extended over all interacting nuclei. This situation is of particular interest in two cases: in paramagnetic complexes in which the unpaired electrons are not localized in the central ion, and in radicals where the unpaired electron has a density distribution function extended over more than one atom. The nature and characteristics of this multiple

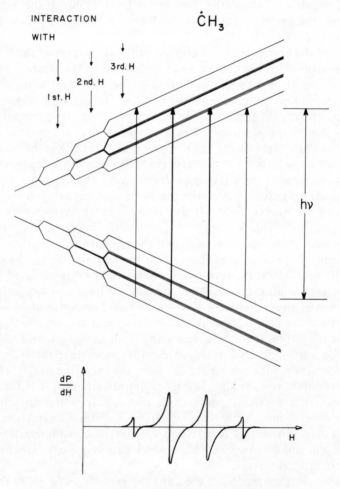

FIG. 10-3 Paramagentic resonance of methyl radical.

interaction may be understood if one considers the simplest case of an electron interacting, with equal hyperfine coupling, with two protons. Since the nuclear spin of a proton is $I = \frac{1}{2}$, there will be four possible configurations of the nuclei, according to the values of m_1, i.e., $(\frac{1}{2}, \frac{1}{2})$, $(\frac{1}{2}, -\frac{1}{2})$, $(-\frac{1}{2}, \frac{1}{2})$, $(-\frac{1}{2}, -\frac{1}{2})$. Of these four configurations, the second and third are equivalent, for it has been assumed that the electron would equally interact with both protons. There will then be three distinguishable configurations in the ratio of occurrence 1:2:1, and the spectrum of such a species will show three equally spaced lines in the same ratio of intensity, i.e., 1:2:1. A spectrum of three equally spaced lines would also be found in the case of interaction of an electron with a deuteron, of spin $I = 1$, but in this case the lines would have the same intensity, for there are three configurations $m_I = -1, 0, +1$, in the ratio of occurrence 1:1:1. We see then that hyperfine splitting due to a single nucleus gives a spectrum of lines of the same intensity. In the particular case of protons, equivalent interaction with n nuclei will give rise to a spectrum of $n + 1$ lines, with a distribution of intensities determined by the coefficients of Newton's binomial. Fig. 10-3 shows the hyperfine spectrum of methyl radical.[20] Notice that natural carbon is practically pure ^{12}C with no nuclear magnetic moment; therefore, the spectrum of methyl radical only displays interaction with protons.

10.6 ENDOR (electron nuclear double resonance)

ENDOR (electron nuclear double resonance), a technique developed in the United States in 1956 by Feher, is an ingenious and elegant method that permits the determination of the hyperfine structure with the resolution of nuclear magnetic resonance.[9] In explaining the principle of ENDOR, one may consider the simplest case of an atom having one unpaired electron ($S = \frac{1}{2}$) and a nucleus of spin $I = \frac{1}{2}$. Some restrictive assumptions will be made to simplify the treatment: the electron is an s electron, its g-factor as well as its hyperfine coupling constant A therefore being isotropic; the substance is magnetically dilute, unpaired electrons then being sufficiently apart to involve no appreciable interaction; and the applied magnetic field is of a certain order of magnitude (to be discussed later). When these conditions are fulfilled, the magnetic energy levels of the unpaired electron, illustrated in Fig. 10.4, are given by the Hamiltonian

$$\mathcal{H} = g\mu_B H \cdot S + A S \cdot I - g_N \mu_N H \cdot I \qquad (10\text{-}47)$$

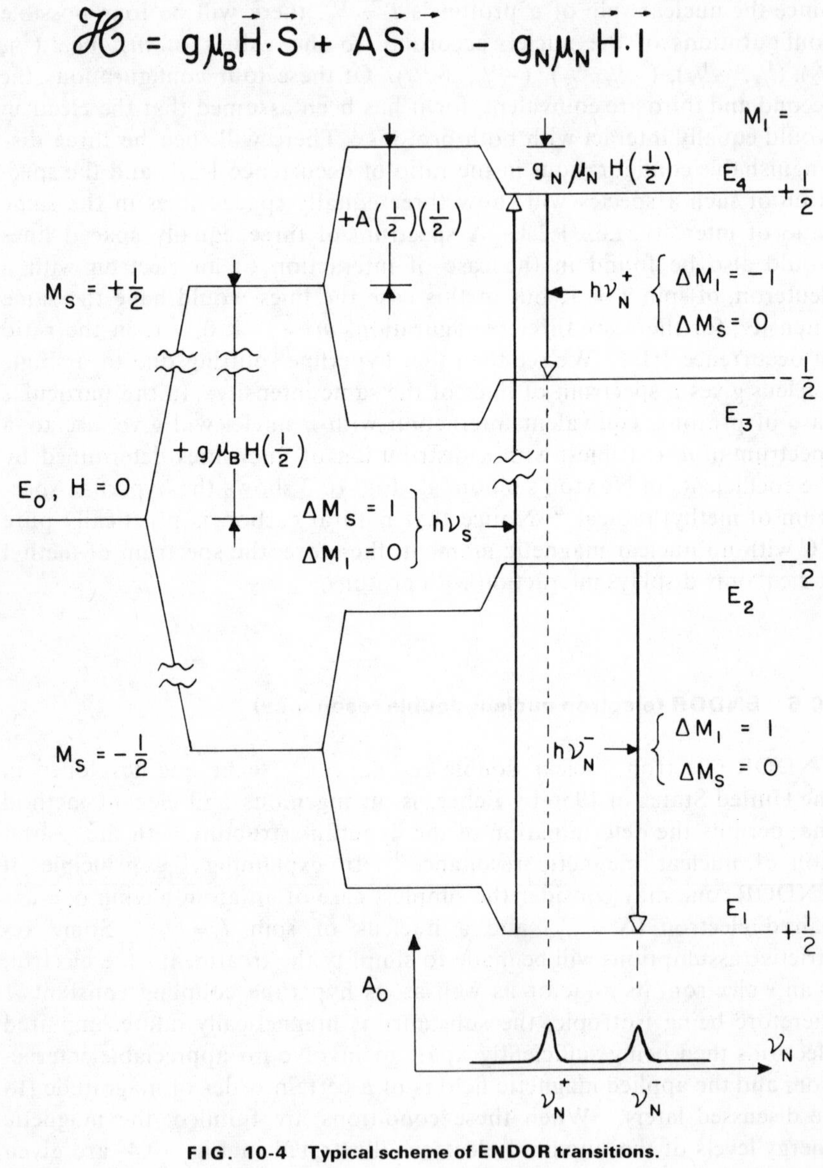

$$\mathcal{H} = g\mu_B \vec{H} \cdot \vec{S} + A\vec{S} \cdot \vec{I} - g_N \mu_N \vec{H} \cdot \vec{I}$$

FIG. 10-4   Typical scheme of ENDOR transitions.

Its eigenvalues are

$$E_4 = \frac{1}{2} g\mu_B H + \frac{1}{4} A - \frac{1}{2} g_N \mu_N H \tag{10-48}$$

$$E_3 = \frac{1}{2} g\mu_B H - \frac{1}{4} A + \frac{1}{2} g_N \mu_N H \tag{10-49}$$

$$E_2 = -\frac{1}{2} g\mu_B H + \frac{1}{4} A + \frac{1}{2} g_N \mu_N H \tag{10-50}$$

$$E_1 = -\frac{1}{2} g\mu_B H - \frac{1}{4} A - \frac{1}{2} g_N \mu_N H \tag{10-51}$$

Allowed transitions are those occurring for $\Delta m_S = \pm 1$ and $\Delta m_I = 0$ (electronic transitions), or $\Delta m_I = \pm 1$ and $\Delta m_S = 0$ (nuclear transitions). Allowed electronic transitions with absorption of energy are then:

(1) E_1 to E_4, by absorption of a photon of energy

$$h\nu_S^+ = E_4 - E_1 = g\mu_B H + \frac{1}{2} A \tag{10-52}$$

and
(2) E_2 to E_3, by absorption of a photon of energy

$$h\nu_S^- = E_3 - E_2 = g\mu_B H - \frac{1}{2} A \tag{10-53}$$

only one of which is indicated in Fig. 10.4. The energy difference between both transitions

$$E_4 - E_1 - (E_3 - E_2) = h(\nu_S^+ - \nu_S^-) = A \tag{10-54}$$

is the hyperfine splitting constant A.

For applied magnetic fields of some thousand oersted, ν_S is of the order of 10 Gc/s and, since ν_S cannot be varied, the spectrum is scanned, at constant frequency ν_S, by varying the magnetic field. The transitions then occur at the values of the magnetic field H_1 and H_2 for which

$$h\nu_S = g\mu_B H_1 + \frac{1}{2} A = g\mu_B H_2 - \frac{1}{2} A \tag{10-55}$$

their difference being

$$g\mu_B \Delta H = A \tag{10-56}$$

where $\Delta H = H_2 - H_1$ is the separation, in oersted, of the hyperfine doublet. There is an important fact that limits the accuracy and resolution of the determination of A by EPR: the width of the absorption lines, that may not only distort but even hide, due to overlapping, any hyperfine structure. The resolution may be improved (by as much as four orders of magnitude) if a combined nuclear and electron resonance experiment is performed, this technique being called ENDOR. It is known that electrons raised to higher levels by resonance absorption of microwave photons are returned to the lower levels by spin-lattice interaction through which they transfer their excess energy to the surroundings. If enough microwave power is fed into the sample at the constant magnetic field of resonance of one of the transitions, the rate at which the electrons are raised becomes greater than the rate at which they can return to the lower level and saturation occurs. Both levels become equally populated and the absorption of microwave power ceases. If a radiofrequency ν_N, variable in the range of nuclear resonance, is now applied to the sample, it is possible to remove the EPR saturation provided the conditions of nuclear resonance are fulfilled. In order to explain this process, saturation of the electronic transition (E_1 to E_4) will be considered. Due to saturation, the population N_4 of level E_4 will be equal to the population N_1 of level E_1 and therefore N_4 will be larger than N_3, the population of level E_3, since in thermal equilibrium and absence of resonance absorption $N_4 \cong N_3 < N_2 \cong N_1$, with N_4 and N_2 only slightly smaller than N_3 and N_1, due to the small nuclear energy difference. If the frequency ν_N is varied within the appropriate range, nuclear resonance will occur at

$$\nu_N = \nu_N^+ = (E_4 - E_3)/h \qquad (10\text{-}57)$$

The population N_4 will then decrease due to the stimulated nuclear transition E_4-to-E_3 and saturation will consequently be removed. This effect is shown by the first peak in the absorption curve of Fig. 10.4. Further increase of ν_N will reach a second nuclear resonance value

$$\nu_N = \nu_N^- = (E_2 - E_1)/h \qquad (10\text{-}58)$$

which will increase the population of E_1, since the saturation value of N_1 will be smaller than N_2 and the frequency ν_N^- will then stimulate the E_2-to-E_1 nuclear transition.

Handling of Eqs. (10-48) to (10-51), (10-57) and (10-58), leads to

$$A = h(\nu_N^+ + \nu_N^-) \qquad (10\text{-}59)$$

and

$$2g_N\mu_N H = h(\nu_N^+ - \nu_N^-) \qquad (10\text{-}60)$$

by eliminating the nuclear and the hyperfine terms. In interpreting Eq. (10-59), the important feature of ENDOR is that the width of the resonance lines is related to those of nuclear magnetic resonance rather than the electron paramagnetic resonance lines. Since nuclear resonance lines are narrower by orders of magnitude, ENDOR makes possible the determination of the hyperfine constant A even when the hyperfine interval is smaller than the EPR line width, in which case conventional EPR experiments could not resolve the hyperfine structure. In addition, it permits the unambiguous assignment of interaction. The nuclear magnetic moment, on the other hand, may be determined with Eq. (10-60) without knowledge of the wave function of the electron through which the measurement is made, approximating the sensitivity of EPR.

It should be noted that the principle behind this latter method is somewhat similar to the Overhauser effect,[17] discovered a year before, which states that under conditions of saturation of the electronic transitions, the difference in population of the nuclear levels is greatly increased. Simultaneous performance of an EPR experiment, therefore, enhances the sensitivity of nuclear magnetic resonance by increasing the intensity of the nuclear transitions.

Cited and general references are listed on page 207, at the end of Chapter 12.

11 *Broadening processes*

11.1 Classification of broadening processes

As it was anticipated in Section 8.9, there are two main categories of broadening: those inherent in the line itself (homogeneous broadening), irrespective of orientation, and those arising from the anisotropic nature of the Zeeman term and the electron dipole-nuclear dipole interaction (inhomogeneous broadening). These effects differ in that some of them are affected by temperature, others by concentration, and others by the state of the sample. The five main effects are:

(1) Spin-lattice interaction, homogeneous
(2) Electron dipole-electron dipole interaction, homogeneous
(3) Exchange interaction, homogeneous
(4) Zeeman anisotropy, inhomogeneous
(5) Electron dipole-nuclear dipole interaction, inhomogeneous

The first effect may be controlled, to a certain extent, by the temperature. The second and third effects are reduced to negligible values by appropriate dilution in a diamagnetic medium. Strictly speaking, exchange interaction is a negative effect, for it actually narrows the lines if the interaction occurs between like paramagnets. Zeeman anisotropy can only be reduced by using lower frequencies. Since it broadens the lines when the paramagnets are randomly oriented, it does not affect measurements performed on single crystals. The fifth effect, due to anisotropic hyperfine splitting of the electron levels by neighboring nuclei, can only be eliminated (as studied in Section 10.4) in liquid samples.

In this chapter, the physical nature of the three first processes will be considered. (The last two processes were discussed in Chapters 9 and 10.)

11.2 Spin-lattice and electronic spin-spin interactions

The effect on the line width of spin-lattice interaction is purely quantum mechanical; it arises from Heisenberg's uncertainty principle. Recall that Heisenberg's uncertainty principle states that it is impossible to determine *precisely* and *simultaneously* the momentum p and the position coordinate q of a particle, or any other pair of conjugate variables. The uncertainty of the determination is given by

$$\delta p \cdot \delta q \sim \hbar \tag{11-1}$$

or, in terms of energy and time

$$\delta E \cdot \delta t \sim \hbar \tag{11-2}$$

If the process of spin-lattice interaction is characterized by a relaxation time τ (defined in Section 8.9), the indetermination in the energy of the transition, $\delta \Delta E$, will be related to τ by

$$\delta \Delta E \cdot \tau \sim \hbar \tag{11-3}$$

Since the energy of the transition is

$$\Delta E = \hbar \omega \tag{11-4}$$

the indetermination in ΔE will result in a *line width* $\Delta \omega$ such that

$$\delta \Delta E = \hbar \Delta \omega \tag{11-5}$$

Elimination of $\delta \Delta E$ between Eqs. (11-3) and (11-5) leaves

$$\tau \cdot \Delta \omega \sim 1 \tag{11-6}$$

A spin-lattice relaxation time T_1 can then be defined by adopting a convention to measure the line width and state

$$T_1 \cdot \Delta \omega = 1 \tag{11-7}$$

if the broadening process arises exclusively from spin-lattice interaction. Using similar arguments, a relaxation time may be assigned to the electronic spin-spin interaction, T_2, so that the line width becomes

$$\Delta\omega = \frac{1}{T_1} + \frac{1}{T_2} \qquad (11\text{-}8)$$

when it is due to both spin-lattice and spin-spin interaction.

The spin-spin relaxation time is often called transverse relaxation time following the terms of nuclear resonance, where both T_1 and T_2 are introduced by means of the so-called Bloch equations, which are briefly discussed in Section 11.3.

Strong interactions are characterized by short relaxation times giving rise to wide lines. Strong spin-lattice interactions are common in paramagnetic ions where the spin-orbit coupling is relatively large. In radicals, the opposite situation occurs and the spin-lattice relaxation times are usually large to the extent that saturation of the lines may be easily achieved. Saturation also leads to broadening, but it may be readily removed by reducing the power fed into the cavity and, if necessary, by increasing the temperature.

The two processes referred to above essentially transform magnetic energy into thermal energy which, in turn, is transferred to the lattice via spin-spin and spin-lattice coupling. The efficiency of the spin-lattice interaction depends on the energy of the spin-orbit coupling. In paramagnetic ions, where spin-orbit coupling is usually large, the spin-lattice relaxation time is of the order of 10^{-8} seconds, whereas in free radicals, where the orbital angular momentum is almost completely quenched, it may be as long as several seconds.

11.3 Bloch equations

Although Bloch's equations[6] were proposed to describe the dynamic magnetic behavior of nuclei interacting among themselves in nuclear magnetic resonance experiments, these equations apply both to nuclei and electrons. They introduce the two relaxation times T_1 and T_2, used in Section 11.2, in a purely phenomenological way.

The basic idea behind Bloch's treatment is that the magnetization of a substance tends exponentially toward its thermal-equilibrium value M_0 due to two well-defined processes of interaction. Suppose that, in our cavity, the electron magnetic moments are precessing about the resultant of the external field and the instantaneous value of the oscillatory field H_1. Turning off the H_1 field will, in general, leave the electron magnetic moment pointing in a direction which will *not* be the direction of the external field H_0. The electron magnetic moment will start to precess about H_0 in a nonequilibrium cone. If the direction of H_0 is taken as the z-axis, the component of the magnetization along z will then increase as the magnetization vector **M** flips toward the z-axis. This process requires the dipole to give up energy to the lattice and is called spin-lattice relaxation. At the same time, the component perpendicular to the z-axis will also change due to a more complex effect. The component of the magnetization perpendicular to the z-axis was given, while H_1 was applied, by the vector sum of all the individual electron magnetic moments along such direction, which were precessing about individual fields arising from the applied ones, H_0 and H_1, and those due to local magnetic interactions. As soon as H_1 is switched off, the coherence of the individual components is lost and $M_\perp$, the component of the magnetization perpendicular to M_z, starts to decay, also exponentially. This process involves spin-spin rather than spin-lattice interaction and is expected to have a different relaxation time, T_2, since the absolute value of the magnetization also changes toward the equilibrium value. Since both relaxations are supposed to be exponential, they may be quantitatively expressed, in a frame rotating with $M_\perp$, by

$$(d/dt) M_z = (1/T_1)(M_0 - M_z) \tag{11-9}$$

$$(d/dt) M_\perp = -(1/T_2) M_\perp \tag{11-10}$$

which are Bloch's equations. Further handling of these equations permits a quantitative evaluation of the line shapes. Although the derivations are out of the range of this book, Pake's monograph gives an excellent account on the subject.

11.4 Exchange narrowing

When unpaired electrons are close enough, it is possible for them to exchange their spin orientation between different atomic or molecular orbitals. The derivation of these effects shall not be presented here. The two-particle model is considered in Van Vleck's book. The many-body problem is treated in Dirac's "Principles of Quantum Mechanics." A comprehensive account of the problem as a whole may be found in Pake's book. The discussion here will be restricted to a qualitative description of the phenomenon.

Exchange interaction is a direct consequence of Heisenberg's uncertainty principle and takes place even when there is negligible chemical binding between the molecules having unpaired electrons. It increases with concentration of unpaired spins and may be reduced to negligible levels by diluting the sample with diamagnetic materials. If the exchange is between similar paramagnets, it narrows the absorption line in the center and broadens it in the low absorption level tails. It is then easily recognized, since it almost invariably leads to Lorentzian shapes instead of the Gaussian shapes produced by the other types of broadening. (The analytical characteristics of these line shapes were given in Section 8.9.) It should be pointed out that, unlike the narrowing observed in the liquid phase—which cancels out the anisotropic hyperfine interaction leaving the contact term—exchange interaction eliminates this latter as well. This is why the stable crystalline radical, 1,1-diphenyl-2-picryl hydrazyl (DPPH) is commonly used as a standard in EPR measurements. In the magnetically dense, pure state, solid DPPH exhibits a very sharp single absorption line very close to the free-electron resonance value, while DPPH solutions show a very well resolved hyperfine splitting due to interaction with the nitrogen nuclei (of spin $I = 1$) and, to a lesser extent, with the hydrogens (of spin $I = \frac{1}{2}$) of the carbon rings.

Cited and general references are listed on page 207, at the end of Chapter 12.

12

The spin Hamiltonian.
III. Crystal field term

12.1 Nature of the crystal and ligand fields

In a general discussion of EPR, it is necessary to consider the influence on the paramagnet of its immediate environment. This environment is characterized by the existence of elements of symmetry, and the effect on the paramagnet is determined by the existence of such elements. Three theories have been developed to account for these facts: the crystal field, the ligand field, and the molecular orbital theories. Indeed, these theories are closely related in that the symmetry properties and requirements remain exactly the same in all three cases. The differences among them arise from assigning to chemical bonding and purely electrostatic effects different relative importance.

In the *crystal field theory,* the paramagnetic ion is considered as subjected to purely electrostatic forces, so that its unpaired electrons may be simply described by atomic orbitals strictly localized in the paramagnetic ion (central ion). The influence of the environment is then reduced to evaluating the form of the electrostatic potential produced by the immediate-neighbor atoms treated as point charges.

In the *ligand field theory,* the electrostatic approach is complemented with the assumption that some chemical bonding does, in fact, exist between the central ion and each one of the nearest-neighbor molecules, now called ligand molecules. In this approach, the unpaired electrons cannot be described in terms of pure atomic orbitals, since the electrons are no longer totally localized in the central ion. In fact, if the covalence of the metal-ligand bond is small, this approach leads, for all practical purposes, to the crystal-field results.

In the case of highly covalent bonding, it is necessary to treat the problem by using the *molecular orbital theory* and eventually describe the electrons by means of hybrid orbitals given by linear combination of atomic orbitals (LCAO), resulting, in turn, from considerations based on the molecular symmetry. To what extent these effects must be taken into account is exemplified in the following two cases.

Consider what happens when a transition ion with an unfilled d-shell— for instance V^{3+}, with two d-electrons—is brought into such a field. If the electrostatic interaction is very strong, it will, in general, break down the coupling between the two orbital angular momenta l_1 and l_2. As a consequence, the total angular momentum L will lose its meaning and each electron will eventually orient in a magnetic field as if the other electron did not exist. This situation is known as *strong field case.* If the electrostatic interaction is not strong enough to break the $l_1 - l_2$ coupling but is still strong enough to break the L-S coupling, the angular momentum L orients itself in the magnetic field, and so does S. Such a situation is called *intermediate field case.* The *weak field case* takes place when the interaction with the ligand field consists of only a higher-order splitting of the levels corresponding to different values of $J = L + S$. The first two cases are common in the first transition series (iron group), due to the d-electrons that are practically unshielded against the ligand field. The weak field case is encountered in the lanthanide ions where the unfilled $4f$-shell is well shielded against the ligand field by the filled $5p$ and $6s$ external electron shells.

Another case in which the ligand field plays a dominant role is that of a paramagnetic ion with only one d-electron, such as Ti^{3+}. Interaction with a ligand field of cubic symmetry removes the five-fold orbital degeneracy of the d-level by splitting it into one two-fold and one three-fold orbitally degenerate levels. The orbital triplet will have the higher energy in a tetrahedral symmetry and the lower energy in an octahedral symmetry. In both cases, the center of gravity of the levels is unchanged relative to that in the presence of a spherical field of equivalent strength. The energy difference between these levels lies in the optical range and accounts for the deep colors so common in compounds of paramagnetic ions.

In the following sections, the nature of these interactions will be studied by applying concepts and results that were developed in Chapter 5.

12.2 Orbital angular functions

In order to apply symmetry arguments to ligand environments and understand their influence on the energy levels of the p-, d-, and f-orbitals, it is advisable to review briefly the geometrical characteristics of the electron wave functions.

It will be remembered that Schroedinger's wave equation of a hydrogen-like atom:

$$\nabla^2 \Psi + \frac{8\mu}{\hbar^2}(E - V)\Psi = 0 \qquad (12\text{-}1)$$

where

$$\nabla^2 = \frac{\partial^2}{\partial x^2} + \frac{\partial^2}{\partial y^2} + \frac{\partial^2}{\partial z^2} \qquad (12\text{-}2)$$

may be put, for simplicity, in spherical coordinates (r, θ, ϕ), as

$$\frac{\partial}{\partial r}\left(r^2 \frac{\partial \Psi}{\partial r}\right) + \frac{1}{\sin \theta}\frac{\partial}{\partial \theta}\left(\sin \theta \frac{\partial \Psi}{\partial \theta}\right) + \frac{1}{\sin^2 \theta}\frac{\partial^2 \Psi}{\partial \phi^2} + \frac{8\mu}{\hbar^2}(E - V)\Psi = 0 \quad (12\text{-}3)$$

in which case the wave function, Ψ, may be written as the product of a function of r alone, another of θ alone, and another of ϕ alone, i.e.,

$$\Psi = R(r) \cdot \Theta(\theta) \cdot \Phi(\phi) \cdot \psi_s \qquad (12\text{-}4)$$

where ψ_s accounts for the spin of the electron. The function $R(r)$ has spherical symmetry and is not affected by operations of symmetry. The function ψ_s accounts for the spin degeneracy which, according to Kramer's theorem (Section 2.7), cannot be removed by an electrostatic field. The discussion may then be restricted to the functions $\Theta(\theta)$ and $\Phi(\phi)$.

In restricting the study to the angular functions of θ and ϕ, we can use the spherical harmonics

$$Y_{l,\pm m}(\theta, \phi) = \Theta(\theta) \cdot \Phi(\phi) \qquad (12\text{-}5)$$

meaning that the probability of finding the electron at a distance between r and $r + dr$ from the nucleus, with its angular coordinates having values between θ and $\theta + d\theta$, and ϕ and $\phi + d\phi$, is

$$\Psi^*\Psi\, d\tau = [R_{n,l}(r)]^2 [Y_{l,\pm m}(\theta,\phi)]^2 r^2 \sin\theta\, dr\, d\theta\, d\phi \qquad (12\text{-}6)$$

The wave function Ψ is, of course, normalized, i.e.,

$$\int \Psi^*\Psi\, d\tau = 1 \qquad (12\text{-}7)$$

extended over all possible values of r, θ, ϕ. It is customary to view this situation by considering the electron as having a spatial distribution whose density at each point is given by

$$\vartheta(r) = \Psi^*\Psi \qquad (12\text{-}8)$$

which is the distribution function used in Sections 10.3 and 10.4.

In the case of s-orbitals, the spherical harmonic reduces to a constant. The symmetry of the cloud is then spherical, and an electron in an s state is only affected by the symmetry of the environment in a higher order of approximation (see Section 12.5).

The angular factors associated with p-electrons may be written, in spherical coordinates:

$$p(m_l = +1) \propto e^{i\phi}\sin\theta \qquad (12\text{-}9)$$

$$p(m_l = 0) \propto \cos\theta \qquad (12\text{-}10)$$

$$p(m_l = -1) \propto e^{-i\phi}\sin\theta \qquad (12\text{-}11)$$

In order to deal with symmetry operations, it is better to replace these functions by linear combinations with a simple expression in orthogonal coordinates. Such linear combinations are, in this case:

$$p_x = \frac{p(+1) + p(-1)}{\sqrt{2}} \propto \sin\theta\cos\phi \propto x \qquad (12\text{-}12)$$

$$p_z = p(0) \propto \cos\theta \propto z \qquad (12\text{-}13)$$

$$p_y = -i\,\frac{p(+1) - p(-1)}{\sqrt{2}} \propto \sin\theta\sin\phi \propto y \qquad (19\text{-}14)$$

Similarly, for d-orbitals, the following linear combinations should be chosen:

$$d_{3z^2 - r^2} = d(0) \propto 3\cos^2\theta - 1 \propto 3z^2 - r^2 \qquad (12\text{-}15)$$

$$d_{xz} = \frac{d(+1) + d(-1)}{\sqrt{2}} \propto \sin\theta\cos\theta\cos\phi \propto xz \qquad (12\text{-}16)$$

$$d_{yz} = -i\,\frac{d(+1) - d(-1)}{\sqrt{2}} \propto \sin\theta\cos\theta\sin\phi \propto yz \qquad (12\text{-}17)$$

$$d_{x^2 - y^2} = \frac{d(+2) + d(-2)}{\sqrt{2}} \propto \sin^2\theta\cos2\phi \propto \sin^2\theta(\cos^2\phi - \sin^2\phi) \propto x^2 - y^2$$

$$(12\text{-}18)$$

$$d_{xy} = -i \frac{d(+2) - d(-2)}{\sqrt{2}} \propto \sin^2\theta \sin 2\phi \propto \sin^2\theta \cos\phi \sin\phi \propto xy \quad (12\text{-}19)$$

Following a similar treatment, the seven f-orbitals are obtained:

$$f_{z^3} \propto 5z^3 - 3r^2 z \qquad \text{for } m_l = 0 \qquad\qquad (12\text{-}20)$$

$$f_{yz^2} \propto y(5z^2 - r^2) \qquad \text{for } m_l = \pm 1 \qquad\quad (12\text{-}21)$$

$$f_{xz^2} \propto x(5z^2 - r^2) \qquad \text{for } m_l = \mp 1 \qquad\quad (12\text{-}22)$$

$$f_{xyz} \propto xyz \qquad\qquad\qquad \text{for } m_l = \pm 2 \qquad\quad (12\text{-}23)$$

$$f_{z(x^2-y^2)} \propto z(x^2 - y^2) \qquad \text{for } m_l = \mp 2 \qquad\quad (12\text{-}23)$$

$$f_{y^3} \propto y(y^2 - 3x^2) \qquad \text{for } m_l = \pm 3 \qquad\quad (12\text{-}24)$$

$$f_{x^3} \propto x(x^2 - 3y^2) \qquad \text{for } m_l = \mp 3 \qquad\quad (12\text{-}25)$$

Fig. 12.1 shows the distribution clouds for $1s$, $2p$, and $3d$ orbitals. It must be understood that these clouds, strictly speaking, extend to $r = \infty$, but for all practical purposes, the integral of Eq. (12-7) converges very rapidly around values of r close to the classical Bohr radii. (A pictorial representation of s, p, d, and f clouds for several values of n, using an ingenious photographic device, has been published by White.[23])

From an analysis of the d-clouds shown in Fig. 12.1, it is a temptation to give a special significance to the z-axis in view of the fact that the $d_{3z^2-r^2}$ cloud has a shape distinctly different from those of the other d-clouds, all of which resemble one another to the extent that they interchange among one another by suitable changes in coordinates. There is not such a special significance of the z-axis. The singular shape of the $d_{3z^2-r^2}$ orbital arises simply from the choice of the linear combination for expressing it in terms of x, y, z. It would be possible to choose linear combinations giving rise to analogous shapes but with their axes along the x- and y-directions, as follows:

$$\frac{3}{2}\, d_{x^2-y^2} - \frac{1}{2}\, d_{3z^2-r^2} \propto 3x^2 - r^2 \propto d_{3x^2-r^2} \qquad (12\text{-}27)$$

$$-\frac{3}{2}\, d_{x^2-y^2} - \frac{1}{2}\, d_{3z^2-r^2} \propto 3y^2 - r^2 \propto d_{3y^2-r^2} \qquad (12\text{-}28)$$

That these modes have actual physical meaning is pointed out by Kauzmann in his book "Quantum Chemistry," by his notation that the moon excites vibrations on the oceans approximating to the $d_{3x^2-r^2}$ and $d_{3y^2-r^2}$ modes with a phase difference of $\pi/2$ between them. This explains the period of the tides, which is 12 hours instead of the approximately 24 hour period of rotation of the earth about the moon.

The angular functions given in this section are not restricted to atomic

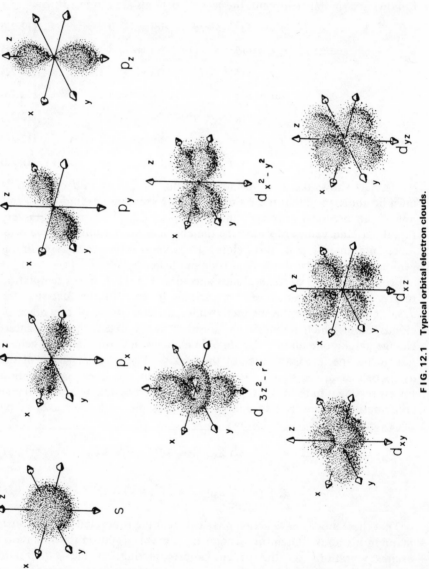

FIG. 12.1   Typical orbital electron clouds.

orbitals of single electrons. When orbit-orbit coupling between electrons occurs, the angular orbital function that describes the resulting total angular momentum **L** has the forms described above. Thus, S, P, D, F, ... ions are assigned s-, p-, d-, f-, ... orbitals. Since two or more unpaired d-electrons filling the d-shell (according to Hund's rules, Sections 2.3 and 2.4) will add their magnetic quantum numbers $m_l = +2, +1, 0, -1, -2$, electronic configurations nd^q will correspond to D terms if $q = 1,4,6,9$, to S terms if $q = 5$, and to F terms if $q = 2,3,7,8$. Fig. 12.2 illustrates this state of affairs. It is left to the reader, as a useful exercise, to prove that f-electrons will in turn give rise to S, F, H, and I terms (information may be found in Section 2.4).

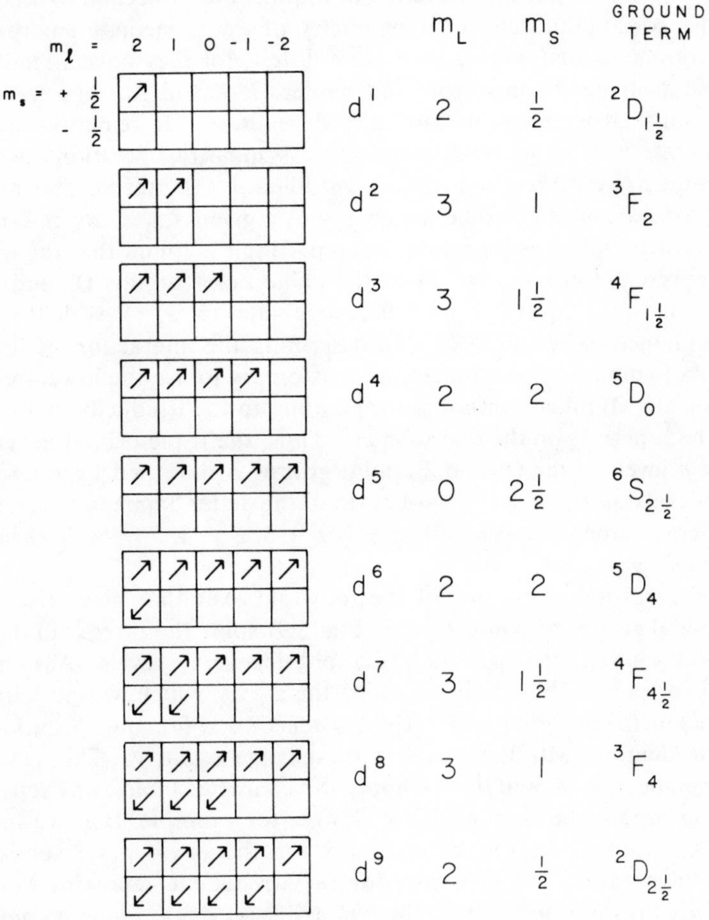

		m_L	m_S	GROUND TERM
d^1		2	$\frac{1}{2}$	$^2D_{1\frac{1}{2}}$
d^2		3	1	$^3F_2$
d^3		3	$1\frac{1}{2}$	$^4F_{1\frac{1}{2}}$
d^4		2	2	$^5D_0$
d^5		0	$2\frac{1}{2}$	$^6S_{2\frac{1}{2}}$
d^6		2	2	$^5D_4$
d^7		3	$1\frac{1}{2}$	$^4F_{4\frac{1}{2}}$
d^8		3	1	$^3F_4$
d^9		2	$\frac{1}{2}$	$^2D_{2\frac{1}{2}}$

FIG. 12.2   Ground terms of the d-ions.

12.3 Removal of orbital degeneracy by the symmetry of the environment

Simple symmetry arguments may be applied to p-orbitals, for their angular functions are functions of x alone, of y alone, and of z alone, i.e.,

$$p_x \propto x \tag{12-12}$$

$$p_y \propto y \tag{12-13}$$

$$p_z \propto z \tag{12-14}$$

These arguments will introduce us to the fascinating, though sometimes cumbersome, subject of interaction of atomic and molecular orbitals with the electrostatic and chemical symmetry of the immediate environment. The arguments that will be used are not new, for they have already been used in dealing with anisotropy in Chapters 7, 9, and 10. In a free atom, the p-subshell is three-fold orbitally degenerate. It remains three-fold degenerate in a cubic field, since some symmetry operations exchange x, y, and z (e.g., three-fold rotations studied in Chapter 5), and no symmetry operation can change the energy of a given state. By referring to the character tables of Appendix V, on p. 218, it is found that the irreducible representations T_{1u} and T_2 of the cubic point groups $\mathbf{O}_h$ and $\mathbf{T}_d$ explicitly state the equivalence of the coordinates (x, y, z), with the meaning explained in Section 5.14. In describing the interaction of the level with the symmetry environment, it is customary to use the lower-case version of the Mulliken symbol corresponding to the irreducible representation that operates on the coordinates. Thus, the triple orbital degeneracy of the p-levels in the $\mathbf{O}_h$ and $\mathbf{T}_d$ point groups is indicated by referring to the p-levels as t_{1u} and t_2. Notice that the letter immediately tells the degeneracy order, i.e., $a, b = 1, e = 2, t = 3, g = 4, h = 5$, *independently of the subscript*.

The distortion along one of the four-fold axes that gives rise to the tetragonal groups of symmetry (see Fig. 5.9) splits the three-fold degenerate p-level into a one-fold and a two-fold degenerate levels. Assume that the distortion has been effected along the z-axis, which will no longer be equivalent to the other two. The existence of a four-fold rotation axis (which identifies with the z-axis) will exchange x and y. The new levels will then be $p_x + p_y$ and p_z. To apply the character tables, any tetragonal point group can be chosen. If we choose, for example, $\mathbf{D}_{4h}$, we find the representation E_u (x, y); the notation for the $(p_x + p_y)$ level is then e_u. The notation for p_z is a_{2u}, for the irreducible representation that explicitly operates on z is A_{2u}. If we now distort the tetragonal configuration along, say, the y-axis, we remove the equivalence between p_x and p_y, meaning that fields of orthorhombic or lower symmetry totally remove the orbital degeneracy of the p-levels, for in these systems no symmetry operation exchanges or mixes x, y, and z.

Interaction of a d-orbital is handled in essentially the same manner. The results are complicated, however, by the fact that free-ion d-levels are five-fold orbitally degenerate. In order to introduce the difficulties gradually, start with an ion of electronic configuration nd^1, where filled subshells are not indicated. Its ground term is then $^2D_{11/2}$ (see Fig. 12.2).

Interaction with a regular octahedral environment is easily studied by resorting to the character table of the O_h point group. The d-level is found to be split into a two-fold orbitally degenerate e_g-level and a three-fold orbitally degenerate t_{2g}-level. In some books, these levels are termed d_γ and d_ϵ. The triple degeneracy of the t_{2g}-level is easily understood: it arises from the equivalence of xy, yz, and xz, which in turn results, in ultimate analysis, from the equivalence of x, y, and z. The degeneracy of the e_g-levels, $d_{3z^2-r^2}$ and $d_{x^2-y^2}$, is not so obvious. However, if one recalls that the orbital angular functions are linear combinations of degenerate solutions (see Section 12.2), the explanation of the two-fold degeneracy of the e_g-levels is clear, as follows.

The $d_{3z^2-r^2}$ function may be put in terms of the angular functions

$$z^2 - y^2 \tag{12-29}$$

$$z^2 - x^2 \tag{12-30}$$

as the linear combination

$$(z^2 - x^2) + (z^2 - y^2) = 2z^2 - x^2 - y^2$$
$$= 3z^2 - (x^2 + y^2 + z^2) = 3z^2 - r^2 \propto d_{3z^2-r^2} \tag{12-31}$$

The second e_g-orbital is

$$d_{x^2-y^2} = x^2 - y^2 \tag{12-18}$$

Any symmetry operation that exchanges x, y, z will similarly exchange Eqs. (12-29), (12-30), and (12-18). A simple choice of the linear combination that will reproduce $d_{3z^2-r^2}$ after a symmetry operation that exchanges x, y, and z will then show that the orbital angular functions $d_{3z^2-r^2}$ and $d_{x^2-y^2}$ have orbital degeneracy in cubic groups.

The obvious question now is which one of the t_{2g}- and e_g-levels will have the higher energy. The answer is illustrated in Fig. 12.3. In the case of an octahedron, where the ligand molecules have the coordinates $(\frac{1}{2}, \frac{1}{2}, 0)$, $(\frac{1}{2}, 0, \frac{1}{2})$, $(0, \frac{1}{2}, \frac{1}{2})$ referred to the edges of the cube showed in solid lines, the e_g-levels experience the strongest interaction, while the t_{2g}-levels experience the weakest interaction. It should be remembered

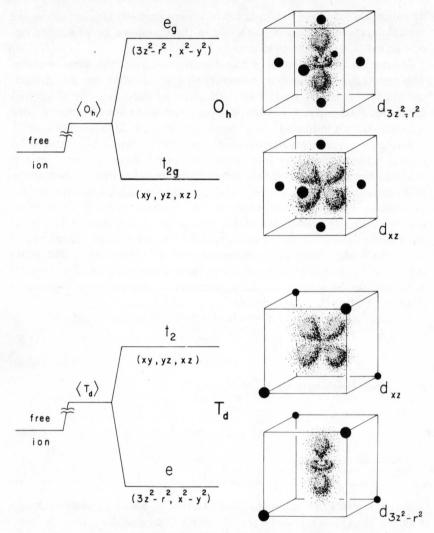

F I G. 12.3   Splitting of the *D*-levels by cubic fields.

that the interaction between an electron in an orbital and a ligand molecule is repulsive, since it arises from interaction between electric charges
of the same sign. Therefore, the potential energy of the e_g-levels must be
higher than the potential energy of the t_{2g}-levels. The two levels are
represented in the same figure, together with five-fold orbitally degenerate
levels corresponding to a spherical field, $\langle \mathbf{O}_h \rangle$, of the same average
strength, and of the free ion. The average potential energy of the electron

in the split levels must be equal to the level in spherical symmetry. The separation between the t_{2g}-level and the spherical-field, five-fold degenerate level must then be equal to two-thirds the separation of the e_g-level to account for orbital degeneracies. The splitting of the d-levels by a regular tetrahedron leads to the opposite situation, as illustrated in Fig. 12.3.

The tetragonal distortion further removes the orbital degeneracy of the d-levels as follows. In lifting the equivalence of z, the $d_{3z^2-r^2}$ and $d_{x^2-y^2}$ are no longer degenerate. Neither are the orbitals d_{xy} and the pair d_{xz}, d_{yz}. There are, then, four d-levels, only one of which (d_{xz}, d_{yz}) is orbitally degenerate. Fig. 12.4 illustrates the splitting of a d^1-ion in two tetragonal fields, compared with the splitting in an octahedral field. The first tetragonal splitting corresponds to a tetragonal octahedron, the second, to a square-planar configuration. Notice the drastic decrease in the energy of the d-levels containing z in their angular functions, which arises from the disappearance of ligand molecules as nearest neighbors along the tetragonal axis. Since both configurations belong to the same point group, $\mathbf{D}_{4h}$, it is clear that the study of the symmetry alone does not suffice for determining the energy of the levels. In order to do this, it is necessary to find an expression for the crystal-field term

$$V_X = -\sum_k e_k \Phi(\mathbf{r}_k) \qquad (9\text{-}11)$$

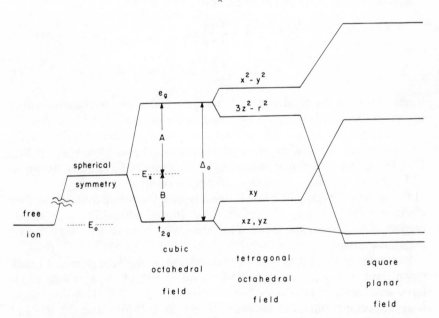

FIG. 12.4   Removal of orbital degeneracies of the D-levels by different crystal fields.

of the Hamiltonian of Eq. (9-1), this being the subject of Section 12.4. It is possible, however, to determine the relative separation of the levels, once their order is known, by performing an ideal experiment. Take, for example, the $e_g - t_{2g}$ splitting by the O_h point group. First bring the free ion, whose energy is E_0 in Fig. 12.3, to the center of a spherical shell of uniformly distributed negative charges whose field is equal to the total field of the octahedral configuration under study. This will raise the ion energy to the value E_s and no splitting of levels will occur, since spherical symmetry is assumed. The energy E_s is greater than E_0 due to the repulsive interaction between the d-electrons and the spherical shell. Now redistribute the charges on the spherical shell so as to place six point charges, each q in magnitude, at the six apices of an octahedron. The value of E_s now becomes an average value, since we know that the two-fold degenerate e_g-level and the three-fold degenerate t_{2g}-level have different energies. Let these energies be $E_s + A$ and $E_s - B$. Since $A + B$ is the splitting Δ_0, it is written, for the ten d-electrons,

$$4(E_s + A) + 6(E_s - B) = 10 E_s \qquad (12\text{-}31)$$

since each orbital may have two electrons of different spin, and

$$A + B = \Delta_0 \qquad (12\text{-}32)$$

Solving Eqs. (12-31) and (12-32) for A and B, we find

$$A = \left(\frac{3}{5}\right)\Delta_0 \qquad (12\text{-}33)$$

$$B = \left(\frac{2}{5}\right)\Delta_0 \qquad (12\text{-}34)$$

which expresses the results, already advanced, that the average potential energy of the split levels must be equal to the energy of the five-fold degenerate level in a spherically symmetrical field of the same total strength. This procedure may be clearly applied to any scheme of levels. In Fig. 12.3, the levels in different symmetry environments have been drawn in scale, assuming that the total field strength is equal in all cases.

The order of magnitude of Δ_0 is, for a typical transition ion, two to four electron volts (i.e., *ca.* 10^4 cm^{-1}) lying in the ultraviolet and visible regions of the optical spectrum. The value $E_s - E_0$ is usually one order of magnitude larger.

Returning to the d-ions under consideration, if the field does not break down the $l_i - l_j$ coupling, the d^q-ions for $q = 1, 4, 6, 8$, all have D-terms and therefore fit into the behavior described so far. However, there is an important difference between the d^1- and d^4-ions and the d^6- and d^8-ions, in that the scheme of levels is inverted. According to the *hole formalism,* which is rigorous, a d^q-ion will behave in the same way as a d^{10-q}-ion, *except* that the interaction energy with the ligand field will have the opposite sign. The physical meaning of this statement is as follows: In ions with less than five d-electrons, there are more empty levels than electrons, and the interaction with the ligand molecules of the electrons under consideration is dominant. If, on the other hand, the ions have more than five d-electrons, there are less empty levels than electrons, and the energy of the configuration is mainly determined by the empty levels. Therefore, the system may be studied as n holes in the d-shell, the holes behaving as positrons. In the special case $q = 5$, $q = 1 - q$, there are as many holes as there are electrons, and the ion is in an S state. No splitting of levels may take place since, for each electron moving from one orbital to another, there is a hole that must be promoted in the opposite direction. An alternative way to view this situation is that since $q = 5$, there being as many holes as electrons, the scheme of this state must be equal to its inverse; therefore, no splitting can occur. Since the problem when q is larger than five may be treated as the interaction of positrons, which are positively charged (instead of the interaction of electrons, which are negatively charged) the energy of the interaction changes in sign and the scheme of levels is inverted. This reduces the complete knowledge of d^q-ions to the study of five cases, namely: $q = 1, \ldots 5$, since the others are derived by inverting the corresponding scheme of levels.

The F-states (see Fig. 12.2) are described by multiparticle wavefunctions analogous to those given in Eqs. (12-20) to (12-26). Since

$$r^2 = x^2 + y^2 + z^2 \tag{12-35}$$

the f-functions may be rewritten and rearranged according to their analytical forms, as follows:

$$xyz \tag{12-36}$$

$$x(y^2 - z^2); \; y(z^2 - x^2); \; z(x^2 - y^2) \tag{12-37}$$

$$x(2x^2 - 3y^2 - 3z^2); \; y(2y^2 - 3z^2 - 3x^2); \; z(2z^2 - 3x^2 - 3y^2) \tag{12-38}$$

The advantage of presenting the f-orbitals in this way becomes obvious when studying the effect of a cubic environment, in which x, y, z are exchanged by symmetry operations. In observing Eqs. (12-36) to (12-38), the conclusion is quickly arrived at that a cubic octahedral environment splits the seven-fold degenerate f-orbitals into one (a_{2u}) nondegenerate level (Eq. 12-36), and two (t_{1u}, t_{2u}) three-fold degenerate levels (Eq. 12-37 and 12-38); for each symmetry operation of the cubic point groups will transform each orbital into itself or, whenever possible, into another orbital of the same row. An analysis of the splittings following the line of argument used in Eqs. (12-31) to (12-34), which is left to the reader, leads to the separations $a_{2u} - t_{2u} = (\frac{6}{5})\Delta_0$ and $t_{2u} - t_{1u} = (\frac{4}{5})\Delta_0$ with the seven-fold orbitally degenerate level in an equivalent spherical field $(\frac{1}{5})\Delta_0$ above or below t_{2u}.

12.4 The crystal field potential

The electrostatic potential about a central ion, due to the ligand molecules, may be expanded in powers of x, y, z and combinations thereof. There are symmetry arguments that impose conditions on the expansion coefficients, with the subsequent simplification of the expression of the potential term $\Phi(\mathbf{r})$ of Eq. (9-11). As usual, starting with a cubic environment reduces the number of terms as follows. Since in any cubic group there are two-fold axes coincident with x, y, and z, or symmetry planes perpendicular to them, or both, the function

$$\Phi(\mathbf{r}) = \Phi(x, y, z) \tag{12-41}$$

must be even in x, y, and z, since it must remain invariant under the operations of symmetry referred to above, which in turn change the sign of each coordinate. Therefore, only even powers of x, y, and z have to be considered in the expansion. Zeroth-order terms in x, y, or z have spherical symmetry and have no bearing in symmetry considerations. The second-order terms may be written

$$C_1 x^2 + C_2 y^2 + C_3 z^2 \tag{12-42}$$

Terms in xy, yz, and xz are odd in each coordinate and cannot appear. Four-fold proper and improper rotations and the four three-fold proper rotations make x, y, and z equivalent. Therefore $C_1 = C_2 = C_3 = C$ and Eq. (12-42) reduces to

$$C(x^2 + y^2 + z^2) = C r^2 \tag{12-43}$$

which is again a spherical function. The next term to consider is the fourth-order, which can be written

$$A_1' x^4 + A_2' y^4 + A_3' z^4 + A_{12}' x^2 y^2 + A_{23}' y^2 z^2 + A_{31}' z^2 x^2 \tag{12-44}$$

Application of the three-fold proper rotations reduces Eq. (12-44) to

$$A'(x^4 + y^4 + z^4) + A_1(x^2 y^2 + y^2 z^2 + z^2 x^2) \tag{12-45}$$

which is invariant under the operations of the cubic group. Higher-order terms, such as $x^2 y^2 z^2, \ldots$, are also invariant. It is usually sufficient to consider the term

$$\Phi = A'(x^4 + y^4 + z^4) \qquad (12\text{-}46)$$

On the basis of analogous arguments, second-order potentials are arrived at

$$\Phi = C_\perp(x^2 + y^2) + C_\parallel(z^2) \qquad (12\text{-}47)$$

and

$$\Phi = C_1 x^2 + C_2 y^2 + C_3 z^2 \qquad (12\text{-}48)$$

that are sufficiently approximate to account for axial and orthorhombic symmetry. Strictly speaking, Eqs. (12-47) and (12-48) should be added, in each case, to the term of Eq. (12-46). However, this latter often represents a negligible effect and hence may be neglected.

In the case of cubic symmetry, the coefficient A' is related to the splitting of the d-levels given in Eq. (12-32) by

$$\Delta_0 = 10\,A' \qquad (12\text{-}49)$$

Theoretical calculations that would take us far afield permit the evaluation of A' for different crystal fields. Thus, one arrives, in the cubic groups, at

$$A'(6) \cong -A'(8) \cong -2A'(4) \qquad (12\text{-}50)$$

where the numbers in parentheses indicate the coordination of the central ion, i.e., 6 in an octahedron, 8 in a cube, and 4 in a tetrahedron. The difference in sign that appears in Eq. (12-50) accounts for the inversion of the scheme of levels referred to above, in dealing with octahedral and tetrahedral configurations.

12.5   The crystal-field spin Hamiltonian

The effect of the crystal field on the Zeeman levels was implicit in the study of the **g**-tensor (Chapter 9). Inclusion of specific terms accounting for the removal of spin degeneracy in paramagnets of spin equal to or larger than one (Figure 12.5) by the crystal field in the Hamiltonian of Eq. (9-15) may be rigorously pursued by applying the perturbation theory (to a higher order in S). Introduced in this section are terms in the effective spin Hamiltonian that fit experimental results and retain their physical meaning of higher-order approximations, without resorting to a rigorous derivation.

The term of the effective spin Hamiltonian that accounts for the removal of spin degeneracy in system of $S \geq 1$ by the crystal field, i.e.,

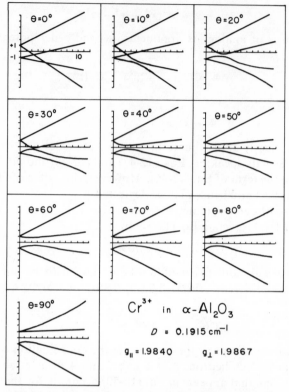

FIG. 12.5 Zero-field and crystal-field splitting for an ion of $S = 3/2$ (Cr^{3+} in α-Al_2O_3) in an axial field, at different orientations. The angle θ is the angle between $g_{\parallel}$ and the magnetic field. The vertical coordinate is divided in units of energy/D. The horizontal coordinate is divided in kilo-oersted. (Reproduced by permission of S. A. Marshall, unpublished material.)

$$\mathbf{S} \cdot \mathbf{\Phi} \cdot \mathbf{S} = [S_x\ S_y\ S_z] \cdot \{\phi_{ij}\} \cdot \begin{bmatrix} S_x \\ S_y \\ S_z \end{bmatrix} \quad (12\text{-}51)$$

was introduced in Eq. (9-15). The diagonal form is

$$[S_x\ S_y\ S_z] \cdot \begin{bmatrix} \phi_1 & 0 & 0 \\ 0 & \phi_2 & 0 \\ 0 & 0 & \phi_3 \end{bmatrix} \cdot \begin{bmatrix} S_x \\ S_y \\ S_z \end{bmatrix} = \phi_1 S_x^2 + \phi_2 S_y^2 + \phi_3 S_z^2 \quad (12\text{-}52)$$

which represents orthorhombic symmetry if

$$\phi_1 \neq \phi_2 \neq \phi_3 \tag{12-53}$$

axial symmetry if

$$\phi_1 = \phi_2 = \phi_\perp, \phi_3 = \phi_\parallel \tag{12-54}$$

and cubic symmetry if

$$\phi_1 = \phi_2 = \phi_3 = \phi \tag{12-55}$$

In the latter case, Eq. (12-51) simplifies to

$$\mathbf{S} \cdot \mathbf{\Phi} \cdot \mathbf{S} = \phi(S_x^2 + S_y^2 + S_z^2) = \phi S(S + 1) \tag{12-56}$$

since

$$S_x^2 + S_y^2 + S_z^2 = S(S + 1) \tag{12-57}$$

and the term accounts for an equal displacement of each preexistent level. Therefore, *a cubic field does not remove, to the second order in S, the spin degeneracy.*

Consideration of Eq. (12-56), which states a condition that must be satisfied by the spin components, reduces the number of independent parameters to two. If one writes

$$\phi_1 S_x^2 + \phi_2 S_y^2 = \frac{1}{2}(\phi_1 + \phi_2)(S_x^2 + S_y^2) + \frac{1}{2}(\phi_1 - \phi_2)(S_x^2 - S_y^2) \tag{12-58}$$

one may define the new parameters

$$E = \frac{1}{2}(\phi_1 - \phi_2) \tag{12-59}$$

$$D = \phi_3 - \frac{1}{2}(\phi_1 + \phi_2) \tag{12-60}$$

Clearly, E vanishes for axial symmetry since $\phi_1 = \phi_2$, and both parameters vanish for cubic symmetry, in view of Eq. (12-55). Now, the average energy of the crystal field is

$$\left(\frac{1}{3}\right)(\phi_1 + \phi_2 + \phi_3) S(S + 1) \tag{12-61}$$

and equally displaces all levels. One may then subtract it from the Hamiltonian, for it will not affect the energy of the transitions. Taking into account the definition of D, Eq. (12-61) may be written as

$$\left[\frac{1}{3} D + \frac{1}{2}(\phi_1 + \phi_2)\right] S(S + 1) \tag{12-62}$$

and subtracted from Eq. (12-52). If Eqs. (12-57), (12-59), and (12-60) are taken into account, the crystal field term, after some algebraic handling, may be written as

$$D\left[S_z^2 - \frac{1}{3} S(S + 1)\right] + E(S_x^2 - S_y^2) \tag{12-63}$$

which is the well-known expression of the crystal field terms widely used in the specialized literature.

In the case of axial symmetry, the term $D[S_z^2 - \frac{1}{3} S(S + 1)]$ directly gives the zero-field splitting, for **H** does not appear in the term. It implies, for example, that for $H = 0$ and $S = 1$, the levels corresponding to different values of $m_S(m_S = 0, \pm 1)$ are separated by $D(1^2 - 0^2) = D$. Analogously, for $S = \frac{3}{2}$, the zero-field splitting is given by $D[(\frac{3}{2})^2 - (\frac{1}{2})^2] = 2D$. The crystal field term gives a zero-field spin degeneracy for levels of $m_S \neq 0$ since the crystal field does not distinguish between $+m_S$ and $-m_S$. This degeneracy is actually lifted by the magnetic field through the Zeeman term giving rise to two levels: $+m_S$ and $-m_S$, which may in turn be split by hyperfine interaction.

Elucidation of the effective spin Hamiltonian (including Zeeman, hyperfine, and crystal field terms) is not always simple, since in orthorhombic symmetries the same reference frame does not always diagonalize all tensors. If it does, then the Hamiltonian simplifies to

$$\mathcal{H} = \mu_B(g_1 S_x H_x + g_2 S_y H_y + g_3 S_z H_z) + A_1 S_x I_x + A_2 S_y I_y + A_3 S_z I_z$$
$$+ D\left[S_z^2 - \frac{1}{3} S(S + 1)\right] + E(S_x^2 - S_y^2) \tag{12-64}$$

If the reference frame does not diagonalize all of the tensors, there *is* a reference frame that diagonalizes the Zeeman and crystal field terms, in which case it is necessary to compute offdiagonal elements in the hyperfine interaction. Axial symmetry simplifies elucidation of EPR spectra, for $E = 0$, and the same reference frame diagonalizes all tensors. The Hamiltonian then reduces to

$$\mathcal{H} = \mu_B(g_\parallel S_z H_z + g_\perp(S_x H_x + S_y H_y)) + A_\parallel S_z I_z + A_\perp(S_z I_z + S_y I_y)$$
$$+ D\left[S_z^2 - \frac{1}{3} S(S + 1)\right] \tag{12-65}$$

Crystal field calculations in axial symmetry are performed after applying the perturbation theory. It is then found that, to the second order in S, the levels of the D-term are[3]

$$D\left(m_S - \frac{1}{2}\right)\left[3 \frac{g_\parallel^2}{g_H^2} \cos^2 \theta - 1\right] \tag{12-66}$$

where m_S, as usual, stands for the spin magnetic quantum number and θ is the angle between the magnetic field and the symmetry axis. The term $(m_S - \frac{1}{2})$ arises from considering a transition between $m_S - 1$ and m_S, i.e.,

$$m_S^2 - (m_S - 1)^2 = 2\left(m_S - \frac{1}{2}\right) \qquad (12\text{-}67)$$

Clearly, there is no crystal field splitting to the second order in S; that is, in m_S^2, when $m_S = \pm\frac{1}{2}$. The variation of the crystal field spectrum, usually known as *fine structure*, then follows a $\left(3\,\dfrac{g_\parallel^2}{g_H^2}\cos^2\theta - 1\right)$ law in axial symmetry. Thus, the splitting of the lines falls to zero and exhibits two maxima, at $\theta = 0°$:

$$2D\left(m_S - \frac{1}{2}\right) \qquad (12\text{-}67)$$

and at $\theta = 90°$:

$$-D\left(m_S - \frac{1}{2}\right) \qquad (12\text{-}68)$$

which permit an immediate evaluation of D. The number of fine structure lines naturally depends on the value of S. In general, each line in a spherically symmetrical field will be split into $2S$ equally spaced lines due to the axial component of the crystal field. In order of magnitude, D is about 1 cm^{-1} varying an order of magnitude in either direction from this estimate. The orthorhombic component E is usually one order of magnitude smaller.

In S-ions like Mn^{2+} and Fe^{3+} (d^5-ions), in which no splitting of the levels to the second order in S occurs, it is necessary to introduce a cubic term

$$\frac{a}{6}\,(S_x^4 + S_y^4 + S_z^4) \qquad (12\text{-}69)$$

which accounts for a fourth-order effect. The parameter a is usually about two orders of magnitude smaller than D. There are, in addition, higher-order effects in D and E that affect all ions. These effects lead to an uneven separation of the fine structure lines. They are treated in detail in the specialized books included in the general references.

When the spin of the nucleus is equal to or larger than one, the nucleus may have an electric quadrupole Q. It is then sometimes necessary to add a term

$$\mathbf{I} \cdot \mathbf{Q} \cdot \mathbf{I} \qquad (12\text{-}70)$$

which may in turn be developed, following analogous lines of argument, as

$$Q'\left[I_z^2 - \frac{1}{3}I(I+1)\right] + Q''(I_x^2 - I_y^2) \tag{12-71}$$

The Q'-term represents axial symmetry and is related to the nuclear electric quadrupole Q by[19]

$$Q' = \frac{3e}{4I(2I-1)}\left(\frac{\partial^2 V}{\partial z^2}\right)Q \tag{12-72}$$

where $\left(\dfrac{\partial^2 V}{\partial z^2}\right)$ is the electric field gradient at the nucleus, along the symmetry axis. The Q''-term accounts for orthorhombic symmetry.

The greatest contribution to the field gradient at the nucleus is usually due to the unpaired electron; the contribution from the ligand molecules (crystal field) is negligible. Some orientations of a nucleus, relative to the quantized unpaired electron, are then more favorable if the nucleus has a quadrupole moment so that there can be an electrostatic interaction that slightly modifies the electronic levels. If one takes the axial case with, for example, $I = 1$, it is found, in the direction of the symmetry axis, that the splitting of levels for each value of m_S, due to the Q'-term, is $Q'[(1^2 - 2) - (0^2 - 2)] = Q'$. The quadrupole effect is even in m_I due to the existence of a mirror plane in the quadrupole, perpendicular to its axis. It should be remembered in this respect that a nuclear quadrupole can be represented by two negative point charges with a double positive charge midway between, no change in configuration remaining after reflection about a perpendicular plane intersecting the quadrupole in the middle point.

The order of magnitude of Q' varies between zero and ca. 10^{-4} cm^{-1}, contributing mainly to broadening of the lines and uneven separation of the spectrum components.

12.6 Spin-orbit coupling

In analyzing the removal of orbital degeneracies, we have only considered the electrostatic forces due to the crystal field and applied symmetry arguments to the solution of the problem. In a more complete treatment, it is necessary to take into consideration the influence of the spin-orbit coupling (Section 2.3) upon the electronic properties of the central ion, which depend, to a large extent, on the order of magnitude of the coupling constant as compared to the splitting of levels due to the crystal field.

This influence arises from the magnetic interaction between the orbital magnetic moment of the electron and its spin magnetic moment, and removes the orbital degeneracy left by an axial field, since in orbitally degenerate states, $L > 0$, and **L** and **S** may couple in more than one way (see Fig. 2.4).

In $3d$-ions, the spin-orbit splitting may often be neglected in view of the much larger crystal-field separations. The spin-orbit and crystal-field splittings are of the same order of magnitude for $4d$- and $5d$-ions. This fact determines compounds of the third transition series ($5d$) to be located midway between the j-j and the L-S coupling schemes.[15] The spin-orbit coupling is, in turn, very large in the $4f$- and $5f$-ions, determining J to be a good quantum number, according to the definition introduced in Section 2.3.

Even though the spin-orbit coupling is small in $3d$-ions, its presence has an important bearing on the magnetic and, to a much lesser extent, optical properties. Magnetically, it determines a departure from the free-electron g-value and characteristic zero-field splittings in EPR measurements.[22] Only in a few cases, however, has it been possible to detect the optical effects of spin-orbit coupling in $3d$-ions.[2]

An excellent example of this state of affairs is provided by the scheme of levels of Cu^{2+} ($3d^9$, $^2D_{21/2}$) in $CuK_2(SO_4)_2(H_2O)_6$, shown in Fig. 12.6. It is shown how the free-ion energy level is lifted to the spherically symmetric crystal-field level, and then split into $t_{2g} + e_g$ due to the cubic field O_h. The orbital degeneracy left by the O_h point group is partially lifted upon introduction of an axial distortion that descends the symmetry of

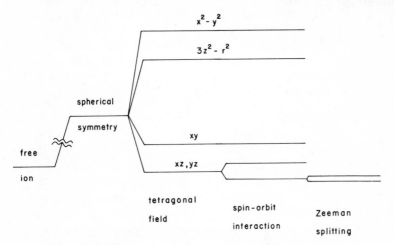

FIG. 12.6   Removal of orbital degeneracy by spin-orbit interaction.

the environment to that of the $\mathbf{D}_{4h}$ point group. The residual degeneracy of the tetragonal e_g-level is finally removed by spin-orbit coupling, leading to five d-levels which have only Kramer's degeneracy. The final splitting of the levels determines the g-values, according to Polder's relations[18]

$$g_\parallel = 2 \left[1 - \frac{4\lambda}{\Delta_\parallel} \right] \qquad (12\text{-}73)$$

$$g_\perp = 2 \left[1 - \frac{\lambda}{\Delta_\perp} \right] \qquad (12\text{-}74)$$

where

$$\Delta_\parallel = E_3 - E_1; \qquad \Delta_\perp = E_{4,5} - E_1 \qquad (12\text{-}75)$$

and $\lambda(Cu^{2+}) = -852$ cm^{-1}. The usefulness of Polder's relations lies in that if they are satisfied (λ and the separation of levels may independently be known from spectroscopic data) they indicate that the electron or hole is localized in the central ion. Otherwise, as in the case of Ag^{2+} ($4d^9$, $^2D_{21/2}$), the hole is not localized in the d-shell but migrates to neighboring ligand molecules.[13]

Application of a magnetic field removes Kramer's degeneracy, as shown in Fig. 12.6. It should be pointed out that due to the large crystal-field splitting, only the lowest level (E_1) is sufficiently populated to allow paramagnetic resonance to occur. Small crystal-field splittings considerably complicate the elucidation of EPR spectra; for in such a case several levels may be sufficiently populated to give rise to measurable resonance absorption.

1. Abragam, A., and Pryce, M. H. L., *Proc. Roy. Soc.* (*London*) **A205**, 135 (1951).
2. Ballhausen, C. J., and Liehr, A. D., *Mol. Phys.* **2**, 123 (1959).
3. Bleaney, B., *Phil. Mag.* **42**, 441 (1951).
4. Bleaney, B., and Ingram, D. J. E., *Proc. Roy. Soc.* (*London*) **A205**, 336 (1951).
5. Bleaney, B., and Stevens, K. W. H., *Rep. Progr. Phys.* **16**, 108 (1953).
6. Bloch, F., *Phys. Rev.* **70**, 460 (1946).
7. Breit, G., and Rabi, I. I., *Phys. Rev.* **38**, 2082 (1931).
8. Cummerow, R. L., and Halliday, D., *Phys. Rev.* **70**, 483 (1946).
9. Feher, G., *Phys. Rev.* **103**, 834 (1956).
10. Johnston, T. S., and Hecht, H. G., *J. Mol. Spectr.* **17**, 98 (1965).
11. Kneubühl, F. K., *J. Chem. Phys.* **33**, 1074 (1960).
12. McConnell, H. M., Heller, C., Cole, T., and Fessendem, F. W., *J. Amer. Chem. Soc.* **82**, 766 (1960).
13. McMillan, J. A., and Smaller, B., *J. Chem. Phys.* **35**, 1698 (1961).
14. Milford, F. J., *Amer. J. Phys.* **28**, 521 (1960).
15. Moffitt, W., Goodman, G. L., Fred, M., and Weinstock, B., *Mol. Phys.* **2**, 109 (1959).
16. Miyagawa, I., Kurita, Y., and Gordy, W., *J. Chem. Phys.* **33**, 1599 (1960).
17. Overhauser, A. W., *Phys. Rev.* **92**, 411 (1953).
18. Polder, D., Physica **9**, 709 (1942).
19. Pound, R. V., *Phys. Rev.* **79**, 685 (1950).
20. Smaller, B., and Matheson, M. S., *J. Chem. Phys.* **28**, 1169 (1958).
21. Townsend, J., Weissman, S. I., and Pake, G. E., *Phys. Rev.* **89**, 606 (1953).
22. Van Vleck, J. H., *J. Chem. Phys.* **7**, 61 (1939).
23. White, H. E., *Phys. Rev.* **37**, 1416 (1931).
24. Zavoisky, E., *J. Phys. USSR* **9**, 211, 245, 447 (1945).

General References (Chapters 8–12)

Roman numerals following the author's name indicate coverage of the following subjects:

(I)   Quantum mechanical fundamentals
(II) Theory of the effective spin Hamiltonian
(III) Experimental techniques
(IV) Application to metal ions
(V)   Application to free radicals
(VI) General applications

Al'tshuler, S. A., and Kozyrev, B. M. (II, III, IV, VI), "Electron Paramagnetic Resonance" (Translated from the Russian), New York, Academic Press, 1964.
Ballhausen, C. J. (I), "Introduction to Ligand Field Theory," New York, McGraw-Hill Book Company, 1962.

Dirac, P. A. M. (I), "Principles of Quantum Mechanics," 4th ed., London, Oxford University Press, 1958.

Eyring, H., Walter, J., and Kimball, G. E. (I), "Quantum Chemistry," New York, John Wiley & Sons, 1944.

Ingram, D. J. E. (III, V), "Free Radicals as Studied by Electron Spin Resonance," London, Buttersworth Publications, 1958.

Ingram, D. J. E. (IV, VI), "Spectroscopy at Radio and Microwave Frequencies," London, Buttersworth Publications, 1955.

Kauzmann, W. (I), "Quantum Chemistry," New York, Academic Press, 1957.

Lebedev, Ya. S., and Voevodskii, V. A., "Atlas of Electron Spin Resonance Spectra," Vol. 1 (translated from the Russian), New York, Consultants Bureau, 1963.

Lebedev, Ya. S., Tikhomirova, N. N., and Voevodskii, V. V., "Atlas of Electron Spin Resonance Spectra," Vol. 2 (translated from the Russian), New York, Consultants Bureau, 1964.

Low, W. (II, VI), "Paramagnetic Resonance in Solids," New York, Academic Press, 1960.

Pake, G. E. (II, III, IV), "Paramagnetic Resonance," New York, W. A. Benjamin, 1962.

Phillips, L. F. (I), "Basic Quantum Chemistry," New York, John Wiley & Sons, 1965.

Poole, C. P., Jr. (III), "Electron Spin Resonance," New York, Interscience Publishers, 1967.

Van Vleck, J. H. (I), "Electric and Magnetic Susceptibilities," London, Oxford University Press, 1932.

APPENDIX I   *Electronic configurations of free atoms*

Atom	Electronic Configuration	Atom	Electronic Configuration
1 H	$1s^1$	11 Na	$(Ne)3s^1$
2 He	$1s^2$	12 Mg	$-3s^2$
3 Li	$(He)2s^1$	13 Al	$-3s^2 3p^1$
4 Be	$-2s^2$	14 Si	$-3s^2 3p^2$
5 B	$-2s^2 2p^1$	15 P	$-3s^2 3p^3$
6 C	$-2s^2 2p^2$	16 S	$-3s^2 3p^4$
7 N	$-2s^2 2p^3$	17 Cl	$-3s^2 3p^5$
8 O	$-2s^2 2p^4$	18 Ar	$-3s^2 3p^6$
9 F	$-2s^2 2p^5$	19 K	$(Ar)4s^1$
10 Ne	$-2s^2 2p^6$	20 Ca	$-4s^2$

Atom	Electronic Configuration	Atom	Electronic Configuration
21 Sc	$-3d^1 4s^2$	63 Eu	$-4f^7 6s^2$
22 Ti	$-3d^2 4s^2$	64 Gd	$-4f^7 5d^1 6s^2$
23 V	$-3d^3 4s^2$	65 Tb	$-4f^9 6s^2$
24 Cr	$-3d^5 4s^1$	66 Dy	$-4f^{10} 6s^2$
25 Mn	$-3d^5 4s^2$	67 Ho	$-4f^{11} 6s^2$
26 Fe	$-3d^6 4s^2$	68 Er	$-4f^{12} 6s^2$
27 Co	$-3d^7 4s^2$	69 Tm	$-4f^{13} 6s^2$
28 Ni	$-3d^8 4s^2$	70 Yb	$-4f^{14} 6s^2$
29 Cu	$-3d^{10} 4s^1$	71 Lu	$-4f^{14} 5d^1 6s^2$
30 Zn	$-3d^{10} 4s^2$	72 Hf	$-4f^{14} 5d^2 6s^2$
31 Ga	$-3d^{10} 4s^2 4p^1$	73 Ta	$-4f^{14} 5d^3 6s^2$
32 Ge	$-3d^{10} 4s^2 4p^2$	74 W	$-4f^{14} 5d^4 6s^2$
33 As	$-3d^{10} 4s^2 4p^3$	75 Re	$-4f^{14} 5d^5 6s^2$
34 Se	$-3d^{10} 4s^2 4p^4$	76 Os	$-4f^{14} 5d^6 6s^2$
35 Br	$-3d^{10} 4s^2 4p^5$	77 Ir	$-4f^{14} 5d^7 6s^2$
36 Kr	$-3d^{10} 4s^2 4p^6$	78 Pt	$-4f^{14} 5d^9 6s^1$
37 Rb	$(Kr) 5s^1$	79 Au$^+$	$-4f^{14} 5d^{10}$
38 Sr	$-5s^2$	79 Au	$(Au^+) 6s^1$
39 Y	$-4d^1 5s^2$	80 Hg	$-6s^2$
40 Zr	$-4d^2 5s^2$	81 Tl	$-6s^2 6p^1$
41 Nb	$-4d^4 5s^1$	82 Pb	$-6s^2 6p^2$
42 Mo	$-4d^5 5s^1$	83 Bi	$-6s^2 6p^3$
43 Tc	$-4d^5 5s^2$	84 Po	$-6s^2 6p^4$
44 Ru	$-4d^7 5s^1$	85 At	$-6s^2 6p^5$
45 Rh	$-4d^8 5s^1$	86 Rn	$-6s^2 6p^6$
46 Pd	$-4d^{10}$	87 Fr	$(Rn) 7s^1$
47 Ag	$-4d^{10} 5s^1$	88 Ra	$-7s^2$
48 Cd	$-4d^{10} 5s^2$	89 Ac	$-6d^1 7s^2$
49 In	$-4d^{10} 5s^2 5p^1$	90 Th	$-6d^2 7s^2$
50 Sn	$-4d^{10} 5s^2 5p^2$	91 Pa	$-5f^2 6d^1 7s^2$
51 Sb	$-4d^{10} 5s^2 5p^3$	92 U	$-5f^3 6d^1 7s^2$
52 Te	$-4d^{10} 5s^2 5p^4$	93 Np	$-5f^4 6d^1 7s^2$
53 I	$-4d^{10} 5s^2 5p^5$	94 Pu	$-5f^6 7s^2$
54 Xe	$-4d^{10} 5s^2 5p^6$	95 Am	$-5f^7 7s^2$
55 Cs	$(Xe) 6s^1$	96 Cm	$-5f^7 6d^1 7s^2$
56 Ba	$-6s^2$	97 Bk	$-5f^8 6d^1 7s^2$
57 La	$-5d^1 6s^2$	98 Cf	$-5f^{10} 7s^2$
58 Ce	$-4f^1 5d^1 6s^2$	99 Es	$-5f^{11} 7s^2$
59 Pr	$-4f^3 6s^2$	100 Fm	$-5f^{12} 7s^2$
60 Nd	$-4f^4 6s^2$	101 Md	$-5f^{13} 7s^2$
61 Pm	$-4f^5 6s^2$	102 No	$-5f^{14} 7s^2$
62 Sm	$-4f^6 6s^2$	103 Lw	$-5f^{14} 6d^1 7s^2$

APPENDIX II   *Magnetic properties of selected ions*

TABLE II.1.   First transition series.

Ion	Term	n^a	$\lambda(cm^{-1})$	$p_{eff.}$ calc.	$p_{eff.}$ obs.
Ti^{3+}	^2D	1	155	1.73	1.8
V^{4+}	^2D	1	250	1.73	1.6–1.7
V^{3+}	^3F	2	210	2.83	2.7
Cr^{3+}	^4F	3	275	3.88	3.8
Cr^{2+}	^5D	4	230	4.90	4.8
Cr^{2+}	^3F	2		2.83	3.3
Mn^{3+}	^5D	4	355	4.90	4.9
Mn^{3+}	^3F	2		2.83	3.2
Mn^{2+}	^6S	5	300	5.92	5.9
Mn^{2+}	^2D	1		1.73	2.2
Fe^{3+}	^6S	5	460	5.92	5.9
Fe^{3+}	^2D	1		1.73	2.4
Fe^{2+}	^5D	4	400	4.90	5.4–5.5
Fe^{2+}	^1S	0		0.00	0.1
Co^{3+}	^5D	4	580	4.90	5.5
Co^{3+}	^1S	0		0.00	0.2
Co^{2+}	^4F	3	515	3.88	4.5–5.1
Co^{2+}	^2D	1		1.73	1.8–2.4
Ni^{2+}	^3F	2	630	2.83	3.3–3.8
Ni^{2+}	^1S	0		0.00	0.0
Cu^{2+}	^2D	1	830	1.73	2.0

$^a n$ = number of unpaired electrons.

TABLE II.2.   Second transition series.

Ion	Term	n^a	$\lambda(cm^{-1})$	$p_{eff.}$ calc.	$p_{eff.}$ obs.
Mo^{5+}	^2D	1	1030	1.73	1.7
Mo^{4+}	^1S	0	950	0.00	0
Mo^{3+}	^4F	3	820	3.88	3.8
Tc^{4+}	^4F	3	1150	3.88	3.8
Ru^{5+}	^4F	3	1500	3.88	3.5
Ru^{4+}	^3F	2	1350	2.83	2.8
Ru^{3+}	^2D	1	1180	1.73	2.1
Rh^{3+}	^1S	0	1360	0.00	0
Pd^{4+}	^1S	0	1830	0.00	0
Pd^{2+}	^1S	0	1460	0.00	0
Ag^{3+}	^1S	0	1930	0.00	0
Ag^{2+}	^2D	1	1840	1.73	1.9

$^a n$ = number of unpaired electrons.

TABLE II.3.   **Third transition series.**

Ion	Term	n^a	$\lambda(cm^{-1})$	$p_{eff.}$ calc.	$p_{eff.}$ obs.
W^{5+}	^2D	1	2500	1.73	1.5
W^{4+}	^1S	0	2100	0.00	0
Re^{5+}	^3F	2	3400	2.83	0.5
Re^{4+}	^4F	3	3000	3.88	3.3
Re^{3+}	^3F	2	2500	2.83	2.1
Re^{2+}	^2D	1	2100	1.73	1.7
Os^{5+}	^4F	3	4300	3.88	3.3
Os^{4+}	^3F	2	4000	2.83	1.3
Os^{3+}	^2D	1	3500	1.73	1.8
Ir^{4+}	^2D	1	5000	1.73	1.7
Ir^{3+}	^1S	0		0.00	0
Pt^{4+}	^1S	0		0.00	0
Pt^{2+}	^1S	0		0.00	0
Au^{3+}	^1S	0		0.00	0

$^a n$ = number of unpaired electrons.

TABLE II.4.   **Lanthanide series.**

Ion	Term	n^a	$\lambda(cm^{-1})$	$p_{eff.}$ calc.	$p_{eff.}$ obs.
Ce^{3+}	$^2F_{21/2}$	1	640	2.56	2.25–2.45
Pr^{3+}	$^3H_4$	2	356	3.62	2.4–2.9
Nd^{3+}	$^4I_{41/2}$	3	270	3.68	3.1–3.5
Sm^{3+}	$^6H_{21/2}$	5	236	1.6	1.4–1.7
Sm^{2+}	$^7F_0$	6	230	3.5	3.6
Eu^{3+}	$^7F_0$	6	227	3.5	3.4
Eu^{2+}	$^8S_{31/2}$	7		7.94	7.9
Gd^{3+}	$^8S_{31/2}$	7		7.94	7.9
Tb^{3+}	$^7F_6$	6	−293	9.7	9.7
Dy^{3+}	$^6H_{21/2}$	5	−364	10.6	9.8–10.5
Ho^{3+}	$^5I_8$	4	−510	10.6	10.4
Er^{3+}	$^4I_{71/2}$	3		9.4	9.2–10.1
Tm^{3+}	$^3H_6$	2		7.6	7.1
Yb^{3+}	$^2F_{31/2}$	1		4.5	3.8

$^a n$ = number of unpaired electrons.

TABLE II.5.   Actinide series.

Ion	Term	n^a	p_{eff} calc.	p_{eff} obs.
U^{5+}	$^2F_{21/2}$	1	2.56	1.7
U^{4+}	$^3H_4$	2	3.62	2.8–3.7
U^{3+}	$^4I_4$	3	3.68	3.5–3.7
Np^{6+}	$^2F_{21/2}$	1	2.56	2.0
Np^{5+}	$^3H_4$	2	3.62	3.2–3.6
Np^{4+}	$^4I_4$	3	3.68	2.9
Pu^{6+}	$^3H_4$	2	3.62	
Pu^{4+}	$^5I_4$	4	2.68	1.4–2.1
Pu^{3+}	$^6H_{21/2}$	5	1.6(?)	0.9–1.2
Am^{3+}	$^7F_0$	6	3.5(?)	1.6
Cm^{3+}	$^8S_{31/2}$	7	7.94	7.9

$^a n$ = number of unpaired electrons.

General References

Figgis, B. N., and Lewis, J., "Progress in Inorganic Chemistry," Vol. 6, F. A. Cotton, editor, New York, Interscience Publishers, 1964.

Figgis, B. N., and Lewis, J., "Technique of Inorganic Chemistry," Vol. 4, H. B. Jonassen and A. Weissberger, editors, New York, Interscience Publishers, 1965.

APPENDIX III   *A caveat for the evaluation of diamagnetic contributions to the susceptibility*

According to Pascal, the molar diamagnetic susceptibility of a pure compound is approximately equal to

$$\chi_M = \sum_i n_i \chi(A)_i + \sum_j n_j \chi(B)_j \qquad \text{(III-1)}$$

where n_i is the number of atoms of kind i and susceptibility $\chi(A)_i$, and n_j the number of structural elements of kind j and susceptibility contribution $\chi(B)_j$. Generally, the susceptibility of a substance may be estimated to within one percent using tabulated values, some of which are displayed in Tables III.1 and III.2.

TABLE III.1.   Selected atomic susceptibilities.

Atom	$\chi(A) \times 10^6$ cgs	Atom	$\chi(A) \times 10^6$ cgs
H	-2.93	O	-4.61
C	-6.00	O(carbonyl)	$+1.73$
C(ring)	-5.76	O(carboxyl)	-3.36
N	-5.57	Cl	-20.1
N(ring)	-4.61	Br	-30.6
N(amide)	-1.54	S	-15.6
N(imide)	-2.11	P	-26.3

TABLE III.2.   Selected structural susceptibilities.

Bond	$\chi(B) \times 10^6$ cgs
C=C	$+5.5$
C=C—C=C	$+10.6$
C≡C	$+0.8$
N=N	$+1.8$
C=N	$+8.2$

Instructional example:

$$\chi(CH_2Cl—CH=CH—CH_3) = 7\chi(H) + 4\chi(C)$$
$$+ \chi(Cl) + \chi(C=C) = -59.1 \times 10^{-6} \text{cgs}$$

Comprehensive tabulations of $\chi(A)$ and $\chi(B)$ may be found in:

Gray, F., and Farquharson, J., *Phil. Mag.* **10,** 191 (1930)

Pacault, A., *Rev. Sci.* **82,** 465 (1944); **86,** 38 (1948)

Pacault, A., Lumbroso-Bader, N., and Hoarau, J., *Cahiers Phys.* **43,** 54 (1949)

Pascal, P., Gallais, F., and Labarre, J. F., *Compt. Rend. Acad. Sci. Paris* **256,** 335 (1963)

A comprehensive tabulation to 1956 of magnetic susceptibilities of diamagnetic and paramagnetic solids is given in:

Föex, G., "Constantes selectionnées; diamagnetisme et paramagnetisme," Paris, Masson et Cie., 1957.

APPENDIX IV   *Measurement and standardization of magnetic fields*

Magnetic resonance of various nuclei may be used to perform precise measurements of the magnetic field. Accuracies of one part in 10^5 are possible with conventional electronic methods. Proton probes (water) are used for measuring magnetic fields between 0.3 and 9 kGauss. Lithium (LiCl) and deuterium (heavy water) are used to cover the 8–24 and 22–60 kGauss ranges.

The NMR (nuclear magnetic resonance) gaussmeter usually consists of the probe and an oscillator-amplifier unit. The resonance may be observed in an oscilloscope. The NMR gaussmeter is used mainly to measure magnetic-field intervals (g-anisotropy, fine and hyperfine structures). Less accurate determinations may be readily performed with a Hall probe, whose principle of operation follows. When a magnetic field is applied to a conductor that is the seat of an electric current of density J, so that the magnetic field H is perpendicular to J, an electric field is generated across the conductor, which is maximum in the direction perpendicular to H and J. This electric field is due to the electric carriers that are displaced toward one side by the magnetic field, and is proportional to the product of H and J. It may then be used, at constant J, for measuring H. Typical Hall probes work at a maximum of 500 mA and reach a sensitivity of 50 mV/kGauss. They are affected by temperature; the voltage output may reduce to one half from 0° to 100°C, but one half of the total drop usually occurs between 0° and 20°C. Semiconducting alloys, such as In-As, are suitable for this purpose. The response is only approximately linear; the accuracy of the Hall probe is of the order of one percent. NMR and Hall gaussmeters are available in the market.

In determining static susceptibilities (discussed in Chapters 6 and 7), it is customary to calibrate the field with a standard paramagnetic compound whose behavior is well known and easily reproducible. Standard paramagnetic compounds commonly used are oxygen (for gaseous susceptibilities) and ferrous ammonium sulphate. The susceptibilities of these substances are given in Table IV.1.

TABLE IV.1.   Magnetic susceptibilities of oxygen and ferrous ammonium sulfate.

Substance	State	$T(°K)$	$\chi \times 10^6$ cgs
Oxygen	gas	293	+3449.0 molar
	liquid	90.1	7699.0
	liquid	70.8	8685.0
	solid, α	54.3	10200.0
	solid, γ	23.7	1760.0
$Fe(NH_4)_2(SO_4)_2 \cdot 6H_2O$	solid	290.2	+32.6 g^{-1}
		77	121.6
		20.2	413
		14.4	547

In EPR experiments, it is an accepted practice to standardize the magnetic field by observing the paramagnetic absorption of pure 1,1-diphenyl-2-pycryl hydrazyl (DPPH). DPPH is a stable radical with one unpaired electron with spin density 0.373 at the α-N and 0.615 at the β-N, the remaining 0.111 being delocalized in the rings:

Pure crystals of DPPH have a single narrow lines of absorption, about two gauss in width, due to exchange interaction. The g-value is 2.0036 ± 0.0003. DPPH 0.1% diluted in single crystals of 1,1-diphenyl-2-pycryl hydrazine shows slightly axial g-tensor with $g_{\parallel} - g_{\perp}$ varying from 0.0010 at room temperature up to 0.0050 at 1.6°K. Dilute solutions show a quintuplet of average interaction with ^{14}N, both α and β, giving $\langle a_N \rangle$ = 8.9 G. The actual interaction with each nitrogen is given by $a_{\alpha} = 9.35 \pm 0.20$ G and $a_{\beta} = 7.85 \pm 0.20$ G. The spin densities in s- and p-states are

$$a_s^2(\alpha) = 0.011 \qquad a_s^2(\beta) = 0.010$$
$$a_p^2(\alpha) = 0.262 \qquad a_p^2(\beta) = 0.605$$

Very dilute solutions (less than 10^{-3} mole/liter) exhibit many hyperfine lines, presumably due to contact terms with nuclei in the rings (H and N). Many substituted DPPH's have been studied; among them, those with $-OCH_3$ and $-F$ instead of a *para*-hydrogen in one of the phenyl groups have g-values equal to 2.000. For a detailed study of DPPH and other stable radicals, the reader is referred to Buchachenko's book, cited at the end of this appendix.

General References

Buchachenko, A. L., "Stable Radicals" (translated from the Russian), New York, Consultants Bureau, 1965.

Föex, G., "Constantes selectionnées; diamagnetisme et paramagnetisme," Paris, Masson et Cie., 1957.

APPENDIX V *Character tables of representative crystallographic point groups*

1. **Triclinic System.**

C_1	E
A	1

C_i	E	i		
A_g	1	1	R_x, R_y, R_z	$x^2, y^2, z^2,$ xy, xz, yz
A_u	1	-1	x, y, z	

2. **Monoclinic System.**

C_s	E	σ_h		
A'	1	1	x, y, R_z	x^2, y^2, z^2, xy
A''	1	-1	z, R_x, R_y	yz, xz

C_2	E	C_2		
A	1	1	z, R_z	x^2, y^2, z^2, xy
B	1	-1	x, y, R_x, R_y	yz, xz

3. **Orthorhombic System.**

C_{2v}	E	C_2	$\sigma_v(xz)$	$\sigma_v'(yz)$		
A_1	1	1	1	1	z	x^2, y^2, z^2
A_2	1	1	-1	-1	R_z	xy
B_1	1	-1	1	-1	x, R_y	xz
B_2	1	-1	-1	1	y, R_x	yz

C_{2h}	E	C_2	i	σ_h		
A_g	1	1	1	1	R_z	x^2, y^2, z^2, xy
B_g	1	-1	1	-1	R_x, R_y	xz, yz
A_u	1	1	-1	-1	z	
B_u	1	-1	-1	1	x, y	

4. **Tetragonal System.**

C_{4v}	E	$2C_4$	C_2	$2\sigma_v$	$2\sigma_d$		
A_1	1	1	1	1	1	z	$x^2 + y^2, z^2$
A_2	1	1	1	-1	-1	R_z	
B_1	1	-1	1	1	-1		$x^2 - y^2$
B_2	1	-1	1	-1	1		xy
E	2	0	-2	0	0	$(x, y)(R_x, R_y)$	(xz, yz)

D_{4h}	E	$2C_4$	C_2	$2C_2'$	$2C_2''$	i	$2S_4$	σ_h	$2\sigma_v$	$2\sigma_d$		
A_{1g}	1	1	1	1	1	1	1	1	1	1		x^2+y^2, z^2
A_{2g}	1	1	1	-1	-1	1	1	1	-1	-1	R_z	
B_{1g}	1	-1	1	1	-1	1	-1	1	1	-1		x^2-y^2
B_{2g}	1	-1	1	-1	1	1	-1	1	-1	1		xy
E_g	2	0	-2	0	0	2	0	-2	0	0	(R_x, R_y)	(xz, yz)
A_{1u}	1	1	1	1	1	-1	-1	-1	-1	-1		
A_{2u}	1	1	1	-1	-1	-1	-1	-1	1	1	z	
B_{1u}	1	-1	1	1	-1	-1	1	-1	-1	1		
B_{2u}	1	-1	1	-1	1	-1	1	-1	1	-1		
E_u	2	0	-2	0	0	-2	0	2	0	0	(x, y)	

5. **Trigonal (Rhombohedral) System.**

C_{3v}	E	$2C_3$	$3\sigma_v$		
A_1	1	1	1	z	x^2+y^2, z^2
A_2	1	1	-1	R_z	
E	2	-1	0	$(x,y)(R_x, R_y)$	$(x^2-y^2, xy)(xz, yz)$

C_{3h}	E	C_3	C_3^2	σ_h	S_3	S_3^5		$\epsilon = \exp(2\pi i/3)$
A'	1	1	1	1	1	1	R_z	x^2+y^2, z^2
E'	$\begin{cases}1 \\ 1\end{cases}$	$\begin{matrix}\epsilon \\ \epsilon^*\end{matrix}$	$\begin{matrix}\epsilon^* \\ \epsilon\end{matrix}$	$\begin{matrix}1 \\ 1\end{matrix}$	$\begin{matrix}\epsilon \\ \epsilon^*\end{matrix}$	$\begin{matrix}\epsilon^* \\ \epsilon\end{matrix}$	(x,y)	(x^2-y^2, xy)
A''	1	1	1	-1	-1	-1	z	
E''	$\begin{cases}1 \\ 1\end{cases}$	$\begin{matrix}\epsilon \\ \epsilon^*\end{matrix}$	$\begin{matrix}\epsilon^* \\ \epsilon\end{matrix}$	$\begin{matrix}-1 \\ -1\end{matrix}$	$\begin{matrix}-\epsilon \\ -\epsilon^*\end{matrix}$	$\begin{matrix}-\epsilon^* \\ -\epsilon\end{matrix}$	(R_x, R_y)	(xz, yz)

6. **Hexagonal System.**

D_{3d}	E	$2C_3$	$3C_2$	i	$2S_6$	$3\sigma_d$		
A_{1g}	1	1	1	1	1	1		x^2+y^2, z^2
A_{2g}	1	1	-1	1	1	-1	R_z	
E_g	2	-1	0	2	-1	0	(R_x, R_y)	(x^2-y^2, xy), (xz, yz)
A_{1u}	1	1	1	-1	-1	-1		
A_{2u}	1	1	-1	-1	-1	1	z	
E_u	2	-1	0	-2	1	0	(x, y)	

D_{6h}	E	$2C_6$	$2C_3$	C_2	$3C_2'$	$3C_2''$	i	$2S_3$	$2S_6$	σ_h	$3\sigma_d$	$3\sigma_v$		
A_{1g}	1	1	1	1	1	1	1	1	1	1	1	1		x^2+y^2, z^2
A_{2g}	1	1	1	1	-1	-1	1	1	1	1	-1	-1	R_z	
B_{1g}	1	-1	1	-1	1	-1	1	-1	1	-1	1	-1		
B_{2g}	1	-1	1	-1	-1	1	1	-1	1	-1	-1	1		
E_{1g}	2	1	-1	-2	0	0	2	1	-1	-2	0	0	(R_x, R_y)	(xz, yz)
E_{2g}	2	-1	-1	2	0	0	2	-1	-1	2	0	0		(x^2-y^2, xy)
A_{1u}	1	1	1	1	1	1	-1	-1	-1	-1	-1	-1		
A_{2u}	1	1	1	1	-1	-1	-1	-1	-1	-1	1	1	z	
B_{1u}	1	-1	1	-1	1	-1	-1	1	-1	1	-1	1		
B_{2u}	1	-1	1	-1	-1	1	-1	1	-1	1	1	-1		
E_{1u}	2	1	-1	-2	0	0	-2	-1	1	2	0	0	(x, y)	
E_{2u}	2	-1	-1	2	0	0	-2	1	1	-2	0	0		

7. Cubic System.

T_d	E	$8C_3$	$3C_2$	$6S_4$	$6\sigma_d$		
A_1	1	1	1	1	1		$x^2+y^2+z^2$
A_2	1	1	1	-1	-1		
E	2	-1	2	0	0		$(2z^2-x^2-y^2, x^2-y^2)$
T_1	3	0	-1	1	-1	(R_x, R_y, R_z)	
T_2	3	0	-1	-1	1	(x, y, z)	(xy, xz, yz)

O_h	E	$8C_3$	$6C_2$	$6C_4$	$3C_2$ $(=C_4^2)$	i	$6S_4$	$8S_6$	$3\sigma_h$	$6\sigma_d$		
A_{1g}	1	1	1	1	1	1	1	1	1	1		$x^2+y^2+z^2$
A_{2g}	1	1	-1	-1	1	1	-1	1	1	-1		
E_g	2	-1	0	0	2	2	0	-1	2	0		$(2z^2-x^2-y^2,$ $x^2-y^2)$
T_{1g}	3	0	-1	1	-1	3	1	0	-1	-1	(R_x, R_y, R_z)	
T_{2g}	3	0	1	-1	-1	3	-1	0	-1	1		(xz, yz, xy)
A_{1u}	1	1	1	1	1	-1	-1	-1	-1	-1		
A_{2u}	1	1	-1	-1	1	-1	1	-1	-1	1		
E_u	2	-1	0	0	2	-2	0	1	-2	0		
T_{1u}	3	0	-1	1	-1	-3	-1	0	1	1	(x, y, z)	
T_{2u}	3	0	1	-1	-1	-3	1	0	1	-1		

A more complete set of character tables for chemically important symmetry groups is included in F. A. Cotton's "Chemical Applications of Group Theory," New York, Interscience Publishers, 1963.

Symbol index* and values of selected physical constants

The values of selected physical constants have been taken from NBS Technical News Bulletin, October 1963. Abbreviations: J, Joule; T, Tesla; C, Coulomb; G, Gauss.

c (velocity of light in a vacuum)
  $2.99792(5 \pm 3) \times 10^{10}$ cm.s^{-1} (CGS);
  $- \times 10^{8}$ m.s.$^{-1}$ (MKSA)

e (electronic charge)
  $1.6021(0 \pm 7) \times 10^{-19}$ C (MKSA);
  $- \times 10^{-20}$ cm$^{1/2}$.g$^{1/2}$ (emu CGS);
  $4.802(98 \pm 20) \times 10^{-10}$ cm$^{3/2}$.g$^{1/2}$.s^{-1}
  (esu CGS)

m_0 (electronic mass)
  $9.109(1 \pm 4) \times 10^{-28}$ g (CGS);
  $- \times 10^{-31}$ kg (MKSA)

h (Planck constant)
  $6.625(6 \pm 5) \times 10^{-27}$ erg.s (CGS);
  $- \times 10^{-34}$ J.s (MKSA)

$\hbar = h/2\pi$
  $1.0545(0 \pm 7) \times 10^{-27}$ erg.s (CGS);
  $- \times 10^{-34}$ J.s (MKSA)

R_∞ (Rydberg's constant)
  $1.097373(1 \pm 3) \times 10^{5}$ cm^{-1} (CGS);
  $- \times 10^{7}$ m^{-1} (MKSA)

μ_B (Bohr magneton)
  $9.273(2 \pm 6) \times 10^{-21}$ erg.G^{-1} (CGS);
  $- \times 10^{-24}$ J.T^{-1} (MKSA)

μ_N (nuclear magneton)
  $5.050(5 \pm 4) \times 10^{-24}$ erg.G^{-1} (CGS);
  $- \times 10^{-27}$ J.T^{-1} (MKSA)

k (Boltzmann constant)
  $1.380(54 \pm 18) \times 10^{-16}$ erg.($^\circ$K)$^{-1}$ (CGS);
  $- \times 10^{-23}$ J.($^\circ$K)$^{-1}$ (MKSA)

N_A (Avogadro constant)
  $6.022(52 \pm 28) \times 10^{23}$ molecule/mole

a contact (Fermi) term, 168; cubic field term, 203; Langevin variable, 35
a_{ij} elements of a square matrix, 65

a_{2u} electron level in a crystal field, 192
a_0 unit vector, 10
a_s s-admixture probability density, 168
A cross section, 101; d-level splitting, 196; isotropic $\mathbf{A}$, 165
$\mathbf{A}$ hyperfine interaction tensor, 155
A_1, A_2, A_3 principal values of $\mathbf{A}$, 165
$A_\parallel, A_\perp$ principal values of axial $\mathbf{A}$, 165
A_i' crystal-field parameters, 198
$A_1, A_2, A_{1u}, A_{2u}, A_{1g}, A_{2g}, A_i', A_i''$ one-dimensional representations, 95
b_{ij} elements of a square matrix, 66
$\mathbf{B}$ dipole-dipole interaction tensor, 168; magnetic induction, 4
B d-level splitting, 196
B_1, B_2, B_3 principal values of $\mathbf{B}$-tensor, 168
$B_\parallel, B_\perp$ principal values of axial $\mathbf{B}$-tensor, 172
B_n normal component of induction $\mathbf{B}$, 4
$B_1, B_2, B_{1u}, B_{2u}, B_{1g}, B_{2g}, B_i', B_i''$ one-dimensional representations, 95
c concentration, 111; velocity of light, 4
c_{ij} elements of a square matrix, 66
C capacitance, 141; Curie constant, 29; torsion constant, 124
C_n n-fold rotation, 69
$\mathbf{C}_n$ rotation point group, 73
$\mathbf{C}_{nv}$ rotation point group, 73
$\mathbf{C}_{nh}$ rotation point group, 73
$\mathbf{C}_{\infty v}$ linear-molecule point group, 74
C_H specific heat at constant magnetic field, 46
C_M specific heat at constant magnetization, 50
C_i crystal-field parameters, 198

*In the list of symbols, the number indicates the page in which the symbol is introduced or defined.

Subject index

1-MONTH